THE MYTH OF THE YEAR

Returning to the Origin of the Druid Calend[illegible]

Helen Benigni
Barbara Carter
Éadhmonn Ua Cuinn

Edited with a foreword by
Miriam Robbins Dexter

University Press of America,® Inc.
Lanham · Boulder · New York · Toronto · Oxford

University Press of America,® Inc.
4501 Forbes Boulevard
Suite 200
Lanham, Maryland 20706
UPA Acquisitions Department (301) 459-3366

PO Box 317
Oxford
OX2 9RU, UK

Printed in the United States of America
British Library Cataloging in Publication Information Available

Library of Congress Control Number: 2003108548
ISBN 0-7618-2572-X (paperback : alk. ppr.)

™ The paper used in this publication meets the minimum requirements of American National Standard for Information Sciences—Permanence of Paper for Printed Library Materials, ANSI Z39.48—1984

In honor of

Marija Gimbutas

(1921-1994)

CONTENTS

FOREWORD

From Neolithic European archaeological remains and from the texts and art of early Indo-European peoples, we find that the goddess appears in a number of animal-forms, particularly those of bird and snake. In *The Myth of the Year*, the authors add intriguing cosmological and astronomical evidence which complements the iconography and myths of bird and snake goddesses, and the divine and heroic in other forms. This evidence mirrors the terrestrial art and the myth to the heavens. We learn that the bird and the snake, along with the celestial waters of the Milky Way, appear together in the skies of the various seasons. Through studying the constellations we can much better understand the wholeness of ancient myth and religion, and the changes reflected in them which occurred when migrating groups both clashed and assimilated with prehistoric peoples.

In the Bronze Age, around 2000 BCE, the migrating Indo-Europeans reached the lands where they would settle, assimilate with indigenous peoples, and form the distinct groups which we know of as the Indo-Iranians, Slavs, Balts, Germanic and Italic peoples, Greeks, Celts, and others. Their languages were imposed upon the indigenous peoples, but particularities of pronunciation and indigenous loan words were incorporated into the new languages, giving each its particular flavor, and eventually each developed into a separate language.

This assimilation is also reflected in the iconographies and early mythical texts of these peoples. Indigenous goddesses were assimilated into the male-dominated pantheons of the Indo-Europeans. Among the Greeks and Celts, goddesses continued to be depicted with the attributes of bird and snake, and sometimes cow. In the sanctuaries of the consort-goddess (once great-goddess), Hera, terracotta snakes have been found. She and Athena, according to Homer, "went forth with steps resembling those of shy doves." In fact, in Homer's *Odyssey*,

Athena flies away, "up into the air, as a bird." A life-sized Athena is to be found in the Acropolis Museum in Athens; snakes encircle her cloak. Coins minted in Athens depicted Athena on the obverse and her owl on the reverse. The magnificent larger-than-life sized statue created by the sculptor Diodalses, the "Crouching Aphrodite" now found in the British Museum, has a snake wound around one of Aphrodite's arms. We can see, in two artifacts in the British Museum, Aphrodite depicted on a clay drinking cup riding a goose, and again Aphrodite with her son Eros as a terracotta figurine, riding a swan – the water birds which in many cultures were associated with the love-goddess. In the Iron Age, Greek goddesses are separated from their animal epiphanies; they are depicted with birds and snakes, rather than being presented as birds and snakes themselves. The grain goddess Demeter is portrayed with both sheaves of wheat and snakes. The Greek traveler, Pausanias, describes the Mistress of Animals, Artemis, holding a torch and two snakes, similar to the Minoan snake-goddess or priestess. However, the earlier Mycenaean Mistress of Animals, the precursor of the classical-age Artemis, is winged; in this earlier era she is one with the bird.

In Ireland, the goddesses retained their earlier avian and viperine forms. The Morrigan was a royston crow, a battlefield goddess, and Nemain and Badv (Badb) were black birds. The mother of the hero Cú Chulainn, Dechtire, also has bird-form. The Celtic Brigid (also called Brigantia and Bride) is the snake, the Queen of the Mound.

In Iron Age Europe and earlier in the Near East, many goddesses bearing bird and snake attributes were demoted to witches and monsters. Iron Age sky maps depict the results of clashing socio-religious systems. On a mythic level, we see the snake now overcome and murdered by the young patriarchal warrior who is the hero and icon of the migrating and invading clashed and assimilated with prehistoric peoples. Sometimes the bird and snake are represented together, as a hybrid. Early on, in the mythologies of the Near East, in the Akkadian Enuma Elish the young warrior hero Marduk killed his grandmother Tiamat, a snaky sea-monster. Elsewhere in the Near East Yahweh "broke to pieces the heads of the sea-monsters in the waters" (Hebrew *Old Testament*, Psalms 74.13) and he also "will strike Leviathan, the fleeing serpent;" then, he "will kill the sea monster, that is in the sea." (Hebrew *Old Testament*, Isaiah 27.1) Among the Indo-Europeans, the Indic warrior-god Indra killed the snaky Danu and her son Vrtra, withholders of the celestial waters, and afterwards Danu and Vrtra lay "like a cow with her calf." (*Rigveda* I.32.8-9) The Greek Apollo killed the Python, the snake who in myth had held oracular Delphi; then

Apollo took over the Oracle for himself. Herakles killed the many-headed Hydra, a water-serpent which lived in a swamp at Lerna. Another Greek hero, Perseus, killed the winged, snaky-haired Gorgon, Medusa. This degeneration of the bird and snake goddess, once the Great Goddess of the Life Continuum, continues as in the Middle Ages angels and knights kill dragons, fire-breathing serpents with wings. The themes of bird, snake, and waters spiral together, on the land, on the sea, and in the skies.

Celestial snakes are compared to cows, adding another layer of richness to the myths. In many groups of peoples, cows are the milk-bringers, the great nourishers. Thus the Egyptian great-goddess and love- goddess, Hathor, is represented as a cow, while the Greek Hera is called the "cow-eyed". In the Irish *Táin Bó Cuailnge*, the Morrígan appears to the hero Cú Chulainn in many forms, including that of a white cow. The celestial Milky Way is mirrored in myth by the Indic Ocean of Milk, the primeval ocean. This ocean was churned by means of the Hindu gods and the demons, working cooperatively to rapidly rotate the head and tail, respectively, of Ananta, the "infinite one", the snake who coiled around Mount Mandara, the *axis mundi*. From this Ocean emerged the elixir of immortality, the amrta, as well as the goddess of love and good fortune, Shrī Lakshmī. The all-nurturing milk gives rise to all good things.

Thus we learn in this book that the starry sky is a cosmic tapestry for the myths of the Neolithic and Bronze Age Goddess of Birth, Death, and Regeneration, the Mistress of Animals, and the Iron Age deities, heroes, and heroines who evolve with the assimilation of the Neolithic Europeans and the Proto-Indo-Europeans.

The Myth of the Year fills both my mind and spirit. It explains, graphically, the intersection of myth and ritual, giving a celestial backdrop for both. It speaks to balances – of lunar and solar cycles, of myth and astronomy. It explains, in a broader way than has been explained before, how the writers of myth are really attempting to chart the world around them, peopling both the planet and the skies with the divine and the heroic, and in so doing, bringing all into harmony with the wheel of the year.

Miriam Robbins Dexter

PREFACE

When a new millennium is dawning and a gyre of time rapidly comes to an end, we either search for new means of measuring time or turn in our need to improve our understanding of time to rediscover ancient calendars. One such ancient calendar is The Sequani Calendar discovered by a group of researchers almost by fate. The researchers, a Celtic Studies group based in the mountains of West Virginia, were in fact in search of an eclectic learning experience when they discovered the ancient text.

The group, now known as The Sequani Celtic Studies Group, got together in the basement of the local college library to enhance their understanding of Celtic Studies by combining disciplines. They felt that the compartmentalizing of disciplines in the modern world had not offered what they termed an eclectic or broader based knowledge of the natural world because it narrowed experience and did not mirror the complex systems of Nature. The group was interested in Celtic Studies because the ancient system of learning proposed by the Druids was a model for them. The Druids combined natural studies, poetry, astronomy, religion and other disciplines to form groups of experts from each field, thus enhancing the picture of the universe to be studied as a whole.

The group embarked on an unknown path of discovery when they received a copy of the Coligny Calendar as it was printed for the Royal Irish Academy in 1926 by a linguist named Eoin MacNeill. MacNeill reworked a reconstruction by Rhys of the original bronze tablets that were discovered in a vineyard near the town of Coligny, France at the turn of the last millennium. The tablets contained vital information for measuring the moon, the stars and the sun, but they were written in a Gaulish tongue that MacNeill could not translate. MacNeill was limited because he approached the calendar from a linguistic point of view and not from an eclectic or Druid standpoint.

With the combined knowledge of several scholars who by now had become trusted companions, the journey of discovery began. The group consisted of Clay and Barbara Carter, an astronomer and astrologer, respectively; Eadhmonn Ua Cuinn, a Celticist and sculptor; Helen Benigni, a writer and mythologer; Mark Butervaugh, a Naturalist and artist; and Tim Krantz, a printmaker. Their first insight, that the calendar was discovered in Coligny, France in the territory of the ancient Celtic tribe of the Sequani, gave the group its name and direction.

A rapid series of discoveries followed. The group discovered that the calendar followed the lunar cycles of the year with amazing accuracy. The names of twelve lunar cycles are listed on the calendar as: Samonios, Dumannios, Rivros, Anagantios, Ogronios, Cutios, Giamonios, Simivisonnios, Equos, Elembivios, Edrinios, and Cantlos. As Barbara began to count lunar returns over 100 years and Clay began to count lunar cycles over longer periods of time, they discovered that for every two and a half years, the calendar reveals an Intercalary Moon, making The Sequani Calendar more accurate than the Julian. This Intercalary Moon is marked on the calendar as an untitled thirteenth moon and might have been used as a holiday for the people; the function of the Intercalary Moon is to keep the lunar and the solar cycles in sync.

To expand the accuracy of The Sequani Calendar further, every 55 years the calendar starts a lunar cycle on one of the four major phases of the moon, returning to its original phase every 220 years. The extra days that the calendar would naturally experience when moving from one phase of the moon to another in 55-year cycles make the Intercalary Moon a longer holiday every 55 years. The calendar begins the lunar cycle on the new moon in the present cycle and marks the first quarter moon as the beginning of the cycle that starts on the Winter Solstice in 2001. Moreover, The Sequani Calendar coincides with other ancient monuments such as Machu Picchu, the Pyramids, and Stonehenge which base their astronomical orientation to the Winter Solstice, a knowledge passed down from the Neolithic peoples.

The Sequani Calendar marks the Winter Solstice or Samonios as the New Year and the beginning of the light half of the year and Summer Solstice or Giamonios as the beginning of the dark half of the year. Each of these Holy Months is a significant holiday designated by two facts. First, the word "Samonios" means the beginning of light and the word "Giamonios" means the beginning of darkness. Second, the holidays are accurately marked in Celtic stone monuments such as Newgrange, Knowth, Dowth, and Stonehenge to coincide with The Sequani Calendar's markings of the Solstices, the Equinoxes, and the lunar cycles.

The Sequani Calendar is therefore an integral part of Celtic culture, and perhaps like other knowledge of astronomy that was passed to the Druids from the Neolithic peoples, the calendar may represent a cornerstone of human achievement which took thousands of years to quantify. Martin Brennan, in his text *The Stones of Time* (Inner Traditions 1994), has discovered two Neolithic calendrical engravings, one on Knowth in Ireland and one on Pola de Alande in Spain, that represent the same cycles on which The Sequani Calendar is based. The fact that the Celtic world which spanned Europe and the British Isles shortly before the turn of the last millennium might have inherited this ancient calendar is not far fetched.

The group's astronomer and astrologer then made the exciting discovery that the beginning of each month of The Sequani Calendar was designated with the appearance of a star of first magnitude, marked PRIN on the calendar, that appears on the Eastern Horizon shortly after sunset. At sunrise, these stars are beginning to set on the Western Horizon. Time is therefore measured at night by the journey of the stars in conjunction with the orbit of the moon. The calendar also looks ahead by marking the next month's star of primary magnitude as it appears in the sky and approaches the beginning of the following month.

The group's mythologer, Helen, correlated the names of the stars and the constellations in which they appear with the goddesses and gods of Celtic mythology. Two months which were undeniably linked to myth served as her basis for exploration into the yearly cycle of Celtic myth and ritual. In the month of Equos, the month celebrating the horse-goddess Epona, Pegasus is clearly visible in the sky, and in the month of Edrinios, the month of sacred passage, the constellation of Eridanus or the River in the Sky is clearly visible. Many of the other months fell into a contextual whole, as research is available to correlate the eight major holidays in Celtic religion with the stars.

The next piece to the puzzle fell into place when the folklore or people's holidays still celebrated in parts of the British Isles today became a vital source of information. The Oenachs, or holidays of importance still celebrated today by many cultures, are clearly marked in the full moon on the calendar and celebrated over a week's time, followed in the same month with observation of the new moon phase designated as Holy Nights or Druid nights, where the Druids worshipped and the people observed a staying-home time. The Oenachs of The Sequani Calendar are devoted to the worship of solar deities as they follow the pattern of the sun's orbit through the year.

The four Oenachs are the Winter Solstice, the Vernal Equinox, the Summer Solstice, and the Autumnal Equinox. On the circle of Neolithic stones, they are North, East, South and West, respectively. In the Neolithic monuments of the Brú na Bóinne in Ireland, Knowth is aligned East-West for the Equinox celebrations of the Oenachs, Dowth is aligned South-West for the sunset of Winter Solstice, and Newgrange is aligned South-East for the Winter Solstice sunrise Oenach. In Britain, Stonehenge is also aligned to the Winter Solstice Oenach when the rays of the sun rise over the heel stone.

The Druid's Holy Nights or stay-home times marked on The Sequani Calendar are discovered in the third quarter phase of the moon and marked in each month as ATENOVX. This is when the Holy Nights begin. In the waning of the new moon phase, the Holy Nights are celebrated. These are fifty and forty-five days from the Oenachs and show their dates to be February 12th, May 12th, and October 12th. (Note that Gregory XIII took ten days out of the Julian Calendar in 1582 calling October 5th, October 15th.)

The Holy Nights form a rectangle outside the circle of Oenachs. Celebrated near the new moon, these Holy Nights are Imbolc, Belteine, Lugnasad, and Samhuin, respectively. Again, certain Neolithic monuments have sacred paths marking the entrance to the circle on a particular Holy Night. These nights are often extended over a seven-night period marked on the calendar as such. The calendar also marks these important nights to carry in two observances each devoted to an aspect of the Holy Night. For example, Samhuin is observed as the Warrior's Samhuin from November 4th to November 12th in Edrinios and the Holy Nights in the new moon of the following the month of Cantlos from December 4th to December 11th.

The group was well on its way to decoding the other markings on The Sequani Calendar as they realized that the lunar cycles, the solar cycles, and the Holy Nights are all decorated by the constellations in a varied and highly accurate picture of celestial meanderings charted meticulously on the calendar. For example, in the Winter Constellations at the peak time of the Winter Solstice in the month of Samonios, there are nine constellations visible, a spectacular setting for the TRINOSAM SINDIV on January 7th. Likewise, in the Summer

Constellations, the Summer Triangle formed by the stars of primary magnitude of the two months leading up to the Summer Solstice in the month of Giamonios, are highly visible for the Holy Night of Midsummer. On the calendar, TIO stars in patterns of three, six and nine most probably forming triangles, sacred patterns of three, are also clearly marked.

When the Sequani Celtic Studies Group realized that this knowledge should have some practical application and use for the public, their three artists, Mark, Eadhmonn, and Tim began plans to silk screen a working calendar incorporating the astronomy, astrology and mythology into the twelve lunar cycles of Samonios, Dumannios, Rivros, Anagantios, Ogronios, Cutios, Giamonios, Simivisonnios, Equos, Elembivios, Edrinios, Cantlos and the Intercalary Moon. The relevance of The Sequani Calendar became more and more evident as the artists began to visualize the natural symbols that would represent each lunar cycle.

The mythology easily matched occurrences in the agricultural cycles, the natural cycles of plants and animals, and in the familiar spiritual cycles and holidays of European cultures. For instance, the agricultural cycles of Lugnasad and the Harvest Festivals on The Sequani Calendar coincide with the harvest festivals already celebrated in many parts of the British Isles and the Continent. The Sacred Snake days of Rivros coincide with St. Patrick's Day and the harvesting of the mistletoe coincides with mid-summer Holy Nights. Most importantly, the holidays of Anagantios and Ogronios coincide with the sacrifice and resurrection of the Christian Easter.

This ancient text is a familiar pattern re-visited. Its accuracy is astounding and its wealth of detail far outweighs our present calendars. It elevates our understanding of time into what the Druids expressed in their script on the ancient monuments in spirals and triskeles, shapes that reflect the eternal patterns of Nature. It lifts us out of the flat two-dimensional perception of time as arbitrary markings on a rectangular surface to a three-dimensional vision of time which spirals to infinity. The Sequani Calendar does this by incorporating the solar, lunar, stellar and natural patterns with the human cycle. With this comes a deeper spiritual vision of our relation to the natural world. It is our opportunity to reach for the stars and use what is within and without.

Helen Benigni
Elkins, West Virginia, 1999

Acknowledgements

Before we launch into the study of the calendar, the authors of the text would like to give thanks for the help they received with the project. Our first acknowledgement is to the stars which were always our guides. We next acknowledge the scholars who supported our efforts and gave us strength: Miriam Robbins Dexter, Morgan Llywelyn, Anne Ross, Peter Berresford Ellis, and J.P. Mallory. Palden Jenkins and Colin Wareham deserve special thanks for their knowledge, and, of course, our friends and family whose very wonder at the world so often encouraged us to look from whence we came have our eternal gratitude. The final touches of gratitude came to us through hands of our artist, Mark Butervaugh, and our layout and design expert, Kathy Doig.

CHAPTER ONE

The Sequani Calendar and the Sacred Calendar of Eleusis

Comparative mythology and the study of culture has been in need of an objective correlative. The calendars of ancient cultures are one such correlative because they measure time by a constant: the patterns of the moon, the stars and the sun. When the myths of the culture are set to these patterns, they are able to tell a story of the myth of the year as it follows the celestial bodies. In conjunction with the celestial bodies and the myths is the pattern of the earth and the agricultural cycle familiar to ancient peoples. By combining all three, we are able to understand mythology as the ancients understood it, as a means of charting the world around them.

Two ancient cultures that serve as sister cultures are the Celtic and the Greek cultures. Both have ancient monuments that chart the celestial bodies as far back as the Neolithic Era retaining many of the rituals and practices of these cultures through symbolic language and artifacts at the sites of their monuments. Both are able to speak the language of the stones when we are able to read them. The Indo-European migrations affected both cultures in terms of their mythologies and in the creation of their languages. With a firm basis in the Indo-European language, both cultures have accessible root meanings to concepts and words. There was also an active exchange between the ancient Celts and the Greeks. Many Druids wrote and spoke Greek, and many Greeks traded and recorded the history and sociology of the Celts.

Most importantly, the calendar of the Celts, or The Sequani Calendar, and the calendar of the Greeks, or the Sacred Calendar of Eleusis, are strikingly similar. The calendars mark lunar and solar time with both cycles in conjunction using an intercalary month to coordinate the cycles of the sun and the moon. The calendars are then successfully able to use lunar cycles for the twelve months of the year. The equinoxes and the solstices are included in the lunar month and the stars are indicated in the calendars as well. The stars are named in Greek mythology to follow a pattern of myths, and Celtic mythology is easily transferable due to the similar archetypes in both cultures.

In the lunar month, both calendars celebrate their people's holidays on the full moon or Oenachs and the priest's Holy Nights on the new moon. The third quarter moon is a significant time in both calendars to begin religious practices. The lunar months are also a microcosm of larger lunar cycles that both calendars follow. Every 55 years the lunar cycle begins on one of the four major phases of the moon. For example, since the middle of the Twentieth Century, the lunar cycle began on a new moon, and in the year 2001, it will begin on the next significant phase of the moon: the first quarter. Many of the Neolithic monuments such Stonehenge with its Aubrey Holes and Knowth with its lunar calendar stone mark these cycles for us.

Both calendars have 55-night cycles that act as a microcosm to that macrocosmic 55-year cycle. In this way, people are able to experience the time in their year and the time of humanity in the lunar cycles. The Sequani Calendar and the Sacred Calendar of Eleusis have a 55-night cycle for one of their most important celebrations: the coming of the fall equinox. Both calendars follow the ellipsis of the sun's orbit and its quickening pace at the equinoxes. The quickening pace of the fall equinox might summon an inherent sense of time coming to a close and the necessity to prepare for the oncoming winter and death itself. Hence, the ancients marked it as a celebration of central importance.

On The Sequani Calendar and the Sacred Calendar of Eleusis, the fall equinox is marked as three-part celebration of the coming of winter and the acceptance of death. The first phase of the 55-night celebration is the Sacred Marriage of the people to the land. The king, as representative of the tribe, marries the goddess of the earth and reaps her harvest. This takes place in the dark half of the lunar cycle of the month of Equos on The Sequani Calendar and in dark half of the lunar cycle of Metageitnion on the Sacred Calendar of Eleusis. The constellations of Pegasus, Equuleus, and Capella mark the celebration and the horse goddesses, such as Epona and Macha in Celtic mythology and Demeter Erinys in Greek mythology represent the celebration.

The second phase of the 55-night celebration, the entire lunar month of Elembivios on The Sequani Calendar and the month of Boedromion on the Sacred Calendar of Eleusis, is set aside as a time to accept the oncoming winter and the end of the harvest as well as the end of one's own life. The seeds of the harvest are a symbol of the potential each person has to accept the inevitability of his or her death. They serve as an agricultural metaphor to the human experience. Both phases of the lunar cycle celebrate this experience and the stars are guides to the myths that enhance it. The Greek goddess, Demeter, and the Celtic goddesses of the Matrona are the representations of this celebration.

The last phase of the 55-night celebration is the lunar cycle of Edrinios on The Sequani Calendar and the lunar cycle of Pyanopsion on the Sacred Calendar of Eleusis. This celebration is to ensure that the seeds of our labor are safely stored for the winter. It is marked by the Hyades, the constellation of the celebration of the Thesmophoria of Pyanopsion where the priestesses of Demeter store the seeds and bless the fields. The Hyades are a cluster of stars that represent Persephone, the daughter of Demeter. As the full moon passes between the Hyades and the Pleiades, the festival is completed.

The understanding of these calendars is the beginning of understanding the myth of the year and the myths of the Celtic and Greek cultures. It is also our pathway to understanding the natural cycles of time and the cycles of the earth in a context of our own experience. In the pages that follow, let the heavens be your guide to this one piece in the larger puzzle of gaining back our time as it is measured by the sun, the moon, and the stars.

Equos, the ninth lunar cycle of the calendar, and the beginning of the celebration of the Fall Equinox on the Greek and Celtic calendars contains the guiding constellations of Pegasus, Equuleus, and Praesepe in Cancer indicating that the cycle of the Divine Horse begins. In the first half of the lunar cycle, the Goddess as horse is seen in her most powerful equine form. She represents controlled strength, independence, and love of spirit. Like the grain, her power is reined in, controlled and harvested. The seventeen nights of the First Harvest are celebrated and abundant grain is prayed for. As a grain goddess, she is a figure of fertility, wealth and sovereignty. Her horse aspect, an Indo-European addition, secures and increases her power. Fomalhaut in Pisces Austrini, known as the Southern Fish, is her guiding star.

The second half, or the dark phase, of the lunar cycle beginning on the third quarter or waning moon, marks the Holy Nights of Equos. The dawn of the last three nights of the dark moon of Equos is especially noted as the beginning of the ceremonies on The Sequani Calendar. This would be the dawn of the 11th, 12th, and 13th nights of the dark phase of the moon. Praesepe, the star cluster in the constellation of Cancer commonly known as the manger, and the two horses, Pegasus and Equuleus, decorate the sky. Similarly, the dark half of the moon in the month of Metageitnion in the Sacred Calendar of Eleusis is marked as the first phase of the mysteries of Demeter, the grain goddess of Greek mythology.

In both cultures, the Goddess, to ensure a good harvest must be united with the people. The king, as representative of the people, sanctifies this important time. The union is the Sacred Marriage of the king to the land as equine power and agricultural wealth. The agricultural significance of the grain that is harnessed and stored is enhanced by the sexual coupling of grain mother with a male companion. In Celtic mythology there are several references to the king coupling with a white mare to impart her blessings on society.

Geraldus Cambrensis in his "Description of Ireland," relates the ritual in Ulster where the people are gathered together to witness the king-elect act out a union with the mare who is then sacrificed and cooked. He becomes the new king by bathing in the broth, eating the flesh, and drinking the broth. The king and the tribe are deemed potent and virile because they have invoked the equine divinity. The lexical support for this ceremony is evidenced in the name Epomeduos, or Epona, the Gaulish name for the compound of "horse" and the intoxicating broth, "medhu," or mead (Mallory and Adams 278). A Celtic artist depicts Epona in a small bronze figurine with ears of corn in her lap and a dish of corn in her right hand which is held high above her steed as a symbol of power and wealth. (Ross 287).

The Irish horse goddess, Macha, inherits the tripartite functions of the Indo-European Horse Goddess as priestess, warrior, and symbol of wealth and fertility, and it is her last function as a symbol of wealth and fertility that is seen in her marriage to the farmer, Crunnchu. Macha greatly increases the wealth of Crunnchu and becomes pregnant with his children. However, she is compelled to take on a race with King Conchobor's horse at the end of her term of pregnancy due to the boasting of her husband. Even though she wins the race in the form of a horse, she gives birth to twins and curses all the men present for making her race. Before she dies in childbirth, she curses the men of

Ulster with birth "pangs" which incapacitate them in battle for nine generations. Perhaps the myth is a warning to the king-elect to be judicious with his newly found power and wealth to avoid evoking the anger of the Divine Horse.

Other figures in Celtic mythology bring the sovereignty of the land to the people through their Sacred Marriages. Medb of Connacht takes several husbands, each of whom becomes king when marrying her. Medb's name itself means a strong and intoxicating broth or mead. Rhiannon, or Rigantona, is the Great Queen of *The Mabinogion* as Divine Horse. Like Macha, she is so swift that no other horse can overtake her. Finally, Étain Echraide, "horse-riding" is the mate of Midir, the god of the Otherworld mound Bri Leith, and the wife of the King, Eochaid Airem.

In Greek mythology, Demeter Erinys is celebrated in the dark half of this lunar cycle through her Sacred Marriage to Poseidon Hippius or Fomalhaut. This is the beginning of the 55-night celebration of the Eleusinian Mysteries of Demeter and Persephone. Demeter is the goddess of the earth's fruits. She is called "the green one," "the bringer of fruit," and the "the one who fills the barn." As grain goddess, she represents the bringer of seasons. Equos and Metageitnion are lunar cycles of the harvest, and the celebrations in both cultures mark this time by the first event of the changing of seasons: the Sacred Marriage of grain mother to the king. In Greek mythology the King, Poseidon marries Demeter as horse goddess and her daughter, Persephone is married to Pluton, the Underworld god of wealth. Because the two goddesses act as mirror figures of each other, both are abducted by gods associated with horses, darkness, death and winter. The Goddesses, as Spring and Summer, must marry Fall and Winter to complete the agricultural cycle of the seasons.

Demeter is told of the abduction of Persephone by Pluton, and she begins searching for her daughter. She is referred to as Demeter Erinys, the angry one, because of her grief; she roamed the earth for nine days without eating, drinking, or bathing. "But when the tenth enlightening dawn had come, Hecate, with a torch in her hands, met her and spoke to her and told her news" of Persephone's abduction (Richardson 293). The dawn of the lunar cycle brings a breakthrough for Demeter and is marked on both calendars. At this time, Poseidon pursues Demeter to Arcadia where Demeter changes into a mare and grazes with the horses of Oncus at Oncieum near Thelpusa. Not at all discouraged, Poseidon changes himself into a stallion and mates with her. She retreats to a cave and gives birth to twins: Areion, a horse, and Despoena, a girl.

The Horse Goddess in both cultures is a symbol of fertility and maternity witnessed by the fact that she bears Divine Twins from the Sacred Marriage. In Indo-European mythology, the twins, or Asvins, are an element to the Sacred Marriage of the Horse Goddess. Demeter and Macha both give birth to a horse and a child. Macha's twins are called Emain Macha, the prehistoric burial mound in Ireland, and Demeter's twins are Areion and Despoena. Despoena was worshipped as a goddess in Arcadia around Phigala.

The Divine Horse Goddess is an expression of "female procreativity and the cyclical rebirth and death of both plants and mankind. She was the Great Mother and the entire world was her Child. The essential event in those religions was the Sacred Marriage, in which the priestess periodically communed with the realm of the spirits within the earth to renew the agricultural year and the civilized life that grew upon the earth. Her male consort was a vegetative spirit, both her son who grew from the earth and her mate who would abduct her to the fecunding other realm as he possessed her upon his death" (Ruck 38).

Elembivios, the lunar cycle that follows Equos and the Sacred Marriage of the Horse Goddess, is lunar cycle devoted to accepting the oncoming of winter where the nights grow significantly longer and the days shorter. It is the last quarter of the year containing the Autumnal Equinox and is generally thought to be an inauspicious time or time to "claim" and settle affairs before the darkness sets in. In the agricultural cycle, it is time to store the grain from the harvest and pray against catastrophes. Spiritually, it is time to realize that the seeds of our labors, if kept in good faith, will be the basis of our spiritual strength. To guide us through Elembivios, we need protection, vision, spiritual strength, and stored wealth.

Elembivios begins as the evening ecliptic has moved from South to North. The star of primary magnitude, or guiding star of the month, Capella in the constellation of Auriga, appears on the Eastern Horizon with the Hyades. When Mercury is visible in the sky, it has its brightest illumination because of the slant of the ecliptic. Mars is a frequent companion to Mercury either in conjunction or in opposition on the horizon at twilight or dawn. This often happens in the last days of the lunar cycle right after the new moon. Sometimes Mars is straight up in the sky. This spectacular portrait in the sky concludes with the crossing of the river in the sky, Eridanus, which changes from a morning constellation to an evening constellation at this time.

This celebration of the tribes is one of the most important celebrations as the stars indicate. Auriga, the constellation called "the

Keeper of Livestock," is the charioteer from the Otherworld of death and winter. He guides us to spiritual realization of the strength of our own accomplishments by showing us that the seeds of our labors are safe in his chariot for the winter. In Celtic mythology, Auriga is the charioteer and path-maker for Cernunnos. Mercury is his high priest aspect, and Mars is his warrior aspect. In Greek mythology, Auriga is Hermes. The Autumnal Equinox complements Auriga as an assessor of the seeds of the harvest in a public Oenach or "Claim Time" on the full moon. When we have balanced our material and spiritual accounts, we cross the River, Eridanus. The crossing of the River in Celtic and Greek mythology is the symbolic journey of the acceptance of life-in-death.

Elembivios is both an agricultural and a spiritual journey. It is a celebration that has its roots in Neolithic times. The Pregnant Vegetation Goddess of Neolithic sites is part of a mystery where her daughter, the seed, dies each year and is resurrected in the spring. The daughter dies annually and the grain mother mourns her death. In this way, the annual cycle of regeneration and germination would be enacted on a yearly basis before the oncoming winter. A sacrificed pig was used as the symbol of the goddess' daughter to reinforce her connection with the earth and the quickly ripening grain. Three grain types, wheat, barley and millet were often stuffed inside clay figurines of the goddess near her altars (Gimbutas 16).

The male consort of the daughter is the underground keeper of wealth and a spiritual priest, the transgressor of souls. He is portrayed as a harvest aspect of the Vegetation Year God with "a crook across his shoulder or attached to his belt." Although we have lost the priest's name, he is depicted on Minoan seals rising from the bucrania, or sacred horns, with a winged goat and a "daemon" with an animal head who holds a pitcher; he is often portrayed on these seals as the "master of animals" (Gimbutas 145). In the stars of Elembivios, he is Auriga, The Charioteer, holding a goat and three kids on his left shoulder and a bridle and whip in his right hand.

In the west room of the east-west temple complexes of the Minoan religion, the rituals preparing people for winter or death took place. The name of the month of these rituals on Crete and Thera is "Eleusinios" (Nilsson 521). The rituals were performed in rooms that contained cult objects with traces of grain and animal offerings. The main symbol in the artwork of these rooms is the horns or the consecrated bucrania which suggest rites of renewal and rites of passage where the initiate would be re-vitalized by a ceremony of death and resurrection. Marija Gimbutas states: "I believe that rituals that occurred in the dark crypts of the Knossos temple complex relate, on the one hand, to those performed one or two thousand years earlier in the large tomb shrines of Old Europe: Newgrange and Knowth in Ireland, and the Hal Saflieni in Malta. On the other hand, they mirrored those ceremonies enacted in classical times, such as the mysteries of Eleusis, accompanied by music and dance, symbolically imitated death and resurrection" (136).

The rituals that Gimbutas speaks of are marked on the ancient calendars of the Greeks and the Celts. On The Sequani Calendar, the dark half of the month, or the last half of the lunar cycle is marked for the Holy Nights of Elembivios. In the first half of the cycle, the people have celebrated their Oenach on the full moon and after feasting and settling their commercial affairs, the Holy Nights of the second half of the lunar cycle are a time for spiritual balancing before the winter. The first nine nights of the cycle are marked as time to cross the River, Eridanus, in a spiritual quest. The sixth and seventh nights of those days are specifically marked with crosses.

On the Sacred Calendar of Eleusis, the same nine nights, mentioned earlier in the Hymn to Demeter, the Holy Nights of the Eleusinian Mysteries, are preceded by a claim time or taking of the tithes for Eleusis. Official delegations proclaiming a holy truce for the first harvest were sent in the name of the grain goddess (Mylonas 244). Again, the sixth and seventh nights of the dark half of the lunar month of Boedromion are the nights in the Telesterion where the Hiera is shown to the people and the life-in-death vision is complete.

The last nights of the dark moon on both calendars are marked for rest, libations, and rites for the dead. The Winter Constellations appear in the night sky on the tenth day of the dark half of the lunar cycle and winter begins. On The Sequani Calendar, these nights are marked by the mysteries of Mars and Mercury; they are marked with three crosses. On the Sacred Calendar of Eleusis, the rites for the dead are called the Plemochoai. Here, a sacred vessel of liquid is poured from a pitcher to the east and west respectively in a cleft in the earth or chthonic chasm (Kerenyi 141). The mystic formula quite possibly ensures that the stored grain contained in the earth will be fertilized in order that it may resurrect in spring. The east-west directions and the pitcher correspond to the Neolithic details of the myth mentioned above.

The Neolithic rituals and the portrait of the stars in Elembivios are also paralleled in Celtic and Greek mythology. In Celtic mythology, the figure of the Pregnant Vegetation Goddess finds her parallels in the Matrona or the Three Mothers of the Celts found in Gaul and Northern Britain. The Matrona are often depicted as three women seated on a stone bench holding cornucopias, fruits, bread, or an infant. The mother figures, as well as other Celtic goddesses, are part of a sacred triskele which represent the daughter, the mother and the crone, three phases of the feminine archetype. In Gaul, the Matrona are often accompanied by Mercury, Mars, or Epona. Single mother goddesses or goddesses with children playing about their feet are variations of the sacred mother found in abundance in Britain and Gaul.

In Irish mythology, Anu is the mother of the tribes called the Tuatha Dé Danaan and in Welsh mythology, Modron, mother of Mabon, is clearly the divine lineage of the hero. In these stories, it is evident that the matriarchy is an important element in the Iron Age myths. According to Anne Ross: "It is clear then from the literary contexts that, not only were the Celts concerned with the concept of a mother goddess who presided over mortals, but they also visualized the gods themselves as belonging to and being controlled by a great divine mother, nurturer of the gods and of the land with which, like all the Celtic goddesses, she was most particularly concerned" (271).

The Celts inherited a particularly strong Indo-European tradition of portraying the pig and the boar in their Iron Age mythology as a harbinger of death, decay and burial. The boar was adopted as a clan symbol because of its ferocity to protect its own. It is associated with heroes and mortality and the battle against death as well as being associated with the vegetation goddesses and the cycle of crops. Instead of a daughter figure that must be revealed to the mother as a sign of the continuity of the vegetation cycle, the hero must hunt and kill the pig or boar as a posture of defense against mortality and death or hunt the animal to the Otherworld and return in triumph.

In the Irish tales of *Magh Mucrime*, the pig has underworld connections ravaging the land for seven years creating a winter of desolation, and in the Fenian cycle of Irish mythology, Finn and his men are often in pursuit of an Otherworldly pig or boar. The Gaulish pig god "Moccus" is identified with Mercury and in the Welsh *Triads* there are three powerful swineherds, Pryderi, Drystan, and Coll who must follow mysterious pigs from the Otherworld across Wales or Britain. Coll's pig gives birth to wheat and barley in the journey thus associating him with the grain mother. The Welsh son of Modron, Mabon, must journey to the Otherworld or Cáer Loyw and battle the magic boar Twrch Trwyth to marry his lover, Olwen.

The third presence that appears in The Sequani Calendar and in the Sacred Calendar of Eleusis is The Horned God of mythology. As the harvest aspect of the Vegetation Year God in Neolithic culture, he is first seen associated with the bucrania as the master of horned animals and as a protector of the goddesses. However, in Indo-European culture, he is given a more dominant role than that of consort. In Indo-European culture, he is named as high priest. The figure of the priest as "one who evoked the deities of the underworld to assist in protecting the fertility of the crops and similar agricultural pursuits" is one of three defined roles of the priesthood. His name "Pont-dheh-ker" means a high priest who makes a path or bridge to the gods (Mallory and Adams 452).

In Celtic mythology, Cernunnos is the antlered stag-god, a dispenser of prosperity who offers his wisdom and prowess, his strength and masculine powers, fecund and immortalizing as a guide through the onset of winter. Like Mercury and Auriga, he brings the vision of immortality, renewal and resurrection to the peoples of the earth. In his horned, animal aspect, he controls the animals. Beasts bow their heads in obeisance to his horned, black image and humans look to him for guidance in controlling their amassed wealth and stored grains for the winter.

Portrayed on the Gundestrup Cauldron in a Buddhic sitting position, antlered and solemn, Cernunnos in his warrior aspect is able to conquer the darkness in the form of a serpent, a symbol of the chthonic regions of the Otherworld. In one hand, Cernunnos firmly grasps the ram-headed serpent and in the other, he holds a torc, symbol of immortality: Twrch Trwyth. Men who are not able, like Cernunnos, to control their wealth are at the mercy of the serpent. This is most likely an Indo-European tradition of conquering the serpent as the goddess of the year, Hera. Cernunnos, divine ancestor, conquering god of the chthonic worlds, balances, like the balanced light of the Equinox, the power between this world and the Otherworld.

In Gaul, Cernunnos is the great ancestor deity of the Gauls, and in Britain, he is The Horned God of the Brigantes. Like the Gaulish Cernunnos, a stone image in Britain of Cernunnos depicts him as a horned god associated with Mercury; he is holding a purse in one hand and accompanied by snakes. According to Ross, many of the British stone figures are significantly different from their Gaulish counterparts because they are markedly phallic. Ross states that there are four basic ways he is depicted in Britain. He is seen as a naked phallic deity, as a warrior, as Mercury, and as a horned head (201).

The Irish iconographic sources attest to a cult of The Horned God and the Irish mythology of the Iron Age reveals a certain tale in the *Táin Bó Cuailnge* concerning Conall Cernach that accurately identifies

Conall as Cernunnos. According to Ross, the tale is "of a very early and genuine Celtic tradition" (196). In the tale, Conall Cernach, like Cernunnos, is an ancestor deity of the Irish. Conall sets out to rescue the Fraech's wife, sons and cattle who have been carried off by a fierce and terrible tribe. Conall meets a crone who tells him to speak to a herdswoman because she cannot help them. The herdswoman, a diviner, tells Conall that she has heard that he is the chosen one who will save Fraech's family and kill the awful serpent that guards them. Conall attacks the fort of the serpent and the serpent slides into his belt without a struggle.

The detail of the serpent in Conall's belt recalls the belt of the Vegetation Year God in Neolithic culture, and the image of Conall controlling or having the serpent encircle his waist recalls a recurrent image in Celtic mythology. The relationship between the great horned-god of the Celts and the serpent symbolizes the control Cernunnos has of the underworld powers. He is the path-maker and protector for the people.

In Greek mythology, the figure of the path-maker is Hermes or Mercury. In the "Hymn to Demeter," Hermes' task is to bring Persephone back to Demeter in his chariot. Although he is a less ominous figure than Cernunnos, Hermes serves a similar purpose as messenger and path-maker between the forces of the Underworld, Hades, and the Earth, Demeter. Most importantly, he carries the vision of immortality that Demeter is shown in order that she may know that her daughter will be kept safely through the winter months and returned in spring. Likewise, the people of the earth are secure because their seeds will be safely stored for the winter to be used again in the spring.

Hermes is a pre-Olympian god associated with the phallus and the snake. He is said to have stimulated plant growth, and he is the keeper of the flocks. Hermes carries a kerykeion or magic staff that has snakes twisted around it. With his kerykeion, he is able to summon souls from the Underworld. In the mysteries of Eleusis, the kykeon is a drink offered to the initiates before they witness the Hiera or vision of Persephone brought to them in Herme's chariot. According to Gimbutas, Hermes is a psychopomp whose function is concerned with hibernating and regenerating (163). As high priest of the Eleusinian mysteries, his function is to transform and alleviate the people of their fear of winter and death.

The myth of Demeter and Persephone is one of the central myths in Greek mythology that has definite roots in Neolithic culture. Demeter is the aspect of the Pregnant Vegetation Goddess that is associated with the fruits of the earth, the grain and the harvest, and her daughter is the grain maiden and queen of the dead. Demeter's sacred animal, the sow, has Neolithic roots and her daughter's association with sprouts, seeds, and the piglet is also Neolithic. According to Gimbutas, the Neolithic myth of Demeter and Persephone involves the daughter of the grain mother descending to the Underworld to hibernate and live in the Underworld as Queen. Persephone's "torch light quickens the grain: the seed does not die, but continues to live in the underworld" (160). She is portrayed as a queen enthroned, holding a dove, a pomegranate, a torch, and ears of corn.

In the mythology of the Indo-Europeans and the Bronze and Iron Age peoples, the Underworld becomes a place of gloom and the seed or maiden is extinguished instead of hibernating. The mother and daughter are both abducted by gods and mourn their separation. The new moon of Equos in The Sequani Calendar and the new moon of Metageitnion in the Sacred Calendar of Eleusis both mark this event as the Sacred Marriage of the horse to the king. Demeter as Demeter Erinys, the angry one, is abducted by Poseidon, and Persephone is abducted by Hades. Demeter must wander the earth in search of her daughter, and the earth experiences winter because of her grief. The Sacred Marriage is in the first 15 nights of a 55-night cycle marked on the Sacred Calendar of Eleusis.

The search and discovery of Persephone is the next 30 nights of the 55-night cycle covering the light and dark halves of the tenth lunar cycles on both calendars. The entire lunar cycle, entitled Elembivios or Boedromion, is dedicated to the search for and discovery of the seed as it sleeps in the earth. In a spiritual sense, we are searching for life-in-death. The final 10 nights of the 55-night cycle is the light half of the month of Edrinios on The Sequani Calendar and the light half of Pyanopsion on the Sacred Calendar of Eleusis. This last part of the 55-night cycle is to cross into death or winter to the Otherworld or Underworld.

Both calendars seem to keep 5-year cycles in order that the solar and lunar cycles are in sync. The Sequani Calendar starts every 55 years on one of the four major phases of the moon for the same reason. The Aubrey Holes at Stonehenge and the lunar calendar stone at Knowth mark those 55-year lunar cycles. On The Sacred Calendar of Eleusis, these cycles are also marked. Obviously, these cycles are marked to keep accurate time with the sun, moon and the stars. The creation of the 55 nights on the Sacred Calendar of Eleusis might then be seen as a microcosm for these larger cycles of time. In this way, the ancients celebrated the season of one year and the season of life simultaneously. The metaphors expand and the meaning enhances as the natural events, both cosmic and definite, are connected.

For this reason, the entire month of Boedromion is devoted to the rituals of Demeter and Persephone, the symbols of "spring-summer and fall-winter in a cycle of constant renewal" (Gimbutas 161). The rituals begin on the first four nights of the dark half of the lunar cycle when the initiates walk fourteen miles from Athens to Eleusis to cross the bridge over the river to the temple of Demeter where they wait at the sacred well or kallichoron outside the temple. Dancing and the sacrifice of a piglet are part of the ceremonies as well as the ritual cleansing of the body and the soul in fasting and prayer.

In the mythology, Demeter must wander the earth and then come to the omphalos or entrance to the Underworld to mourn. Demeter assumes the guise of a crone and Hecate, the actual goddess who is the crone, attends her. Maidens from the house of Metaneira help Demeter by sheltering her for days, but all is to no avail until Hecate brings Hermes. Hermes, the transgressor of souls, then enters the Underworld to retrieve Persephone and bring her to Demeter. On the Sacred Calendar of Eleusis this day, the fifth day of the dark half of Boedromion is called "Pompe." Torches and a procession of priests approach the Sacred Gate where the initiates, posing as Demeter, wait. They cross the bridge between this world and the Otherworld and the veil is lifted.

In The Sequani Calendar and the mythology of the Celts, this is marked as crossing the River on the sixth and seventh nights of the dark half of Elembivios. On the Sacred Calendar of Eleusis, the initiates enter the temple to experience the Hiera on the same nights of Boedromion. A yellow ribbon is tied on the leg and hand of each initiate to symbolize his or her illumination or vision and each drinks the kykeon. The tying of one leg and one hand is reminiscent of the powerful stance of the Druids where they raise one hand and one foot and close one eye to summon the powers of the Otherworld. Having experienced what Carl Kerenyi calls a "vision beatifica," the people understand the powers of life and death.

Unfortunately, Persephone has tasted of the pomegranate, or the fruits of the dead, and must return to the Underworld to remain underground for the winter. The celebrations on The Sequani Calendar are marked by three crosses and the appearance of the Winter Constellations. The Sacred Calendar of Eleusis marks them as the Plemochoai or rites for the dead. In the "Hymn to Demeter," Hecate becomes the minister and companion of Persephone, and Rhea, the goddess of the earth, becomes the companion to Demeter and asks Demeter to join the families of the gods. On earth, the people are given a sacred calendar in order that they may eternally celebrate the mysteries of life and death in accordance with the stars, the moon, the sun and the cycles of the earth's fruits.

The lunar cycle of Edrinios, the last lunar cycle of the celebration of the 55-night cycle, is named for the constellation of Eridanus which appears on the Eastern Horizon at the beginning of the month. Eridanus, the River, is our spiritual guide through the celebration of this lunar cycle. Here, the moon is celebrated in two distinct phases. The first phase is celebrated in the powerful light of the nights of the full moon itself. On the tenth night of the month, when the full moon passes the Hyades, known as the piglets, the second phase of the celebration begins and lasts until the moon is waning in its third quarter. These are the contemplative and sacred nights of the celebration performed by the priestesses of Persephone who are represented by the constellation of the Hyades.

In the 55-night celebration of the grain harvest, Edrinios is the last part of that celebration. The 55-night cycle begins with the Sacred Marriage of the people to the land as a symbol of our commitment to the harvest; it takes place in the dark half of the lunar cycle of Equos. The second part of the 55-night cycle is the entire month of Elembivios and it is devoted to embracing the oncoming winter through a spiritual awakening to life-in-death and a realization of our human potential. The culmination of that cycle reaches its peak in the Holy Nights of Elembivios where we are given spiritual peace to face death and make it through the oncoming months of winter. The last phase of the 55-night cycle is to ensure that the seeds of our labor are safely stored for the winter; this essential month is Edrinios.

On The Sequani Calendar, Edrinios is marked as Tiocobrextio, a celebration of the tribes. The three days around the full moon are the time of the first phase of the Oenach. The second phase of the Oenach is marked on The Sequani Calendar with three crosses before the moon begins its third quarter; this is when the full moon passes the Hyades. On the Sacred Calendar of Eleusis, the celebration is in the month of Pyanopsion, the eleventh lunar cycle of the Greek calendar. The first phase of celebration is called the Proerosia and the Pyanopsia of the Thesmophoria. The second phase of the celebration on the Greek calendar is the festival of the Thesmophoria itself; it too takes place on the nights before the third quarter.

Both celebrations have their roots in the Neolithic culture of Europe and the Mediterranean. They were the times for the storage and the blessing of the seeds, a sacred activity carried out by the women of the tribes. The Hyades are an apt representation of these festivals as they are the piglets of Persephone, most likely represented by her priestesses. In the night sky, they appear as a cluster or group and their rising signifies the beginning of winter. They are often immersed in a cloud of streamers, as if protected by a veil. They are protected as the seeds of our labors must be protected. Before they had a name in the myths and legends of the Celts and the Greeks, they appeared as a sign in the sky that it was time for the winter seeds to be carefully protected and stored.

Marija Gimbutas describes three types Neolithic temples, one of which she calls the temples dedicated to the Pregnant Vegetation Goddess. In the temples were found masks of pigs, clay figurines of young, mature and crone women, and figurines with grain stuffed inside. The temples housed seeds in underground pits, in offering pits in the corners of the temple courtyard, or in "pithoi" jars. Stored grains were blessed and cared for during the winter months and the women who cared for the temples were part of a Neolithic matri-local culture. According to Gimbutas, the care and maintenance of the Neolithic temples was organized by councils of women with several levels including that of priestess and attendants (Gimbutas 120).

The storage of the sacred grain inherited from the Neolithic peoples continued in the agrarian festivals of the Greek Thesmophoria. In the preparation festivals on the full moon of Pyanopsion, called the Proerosia and the Pyanopsia, mixed grains and beans in earthen jars were dedicated to the goddesses and to Apollo Pythios, a representative of the snake. The ceremonies were known as "sacrifices for the future crops before the annual ploughing and its almost certain connection with the ritual pre-ploughing of the sacred Rharian field at Eleusis" (Dow and Healey 15). The priestesses for the ceremony were chosen from a special group of "chaste" women who lived in a sacred dwelling close to the sanctuary of Eleusis; they might have been both the representations of the Hyades, sisters of the heavenly tribe, and representatives of the Neolithic matriarchy on earth.

In the second phase of the celebration, on the three days that precede the third quarter moon, the actual Thesmophoria took place. Here, the remains of piglets sacrificed to Demeter and Persephone in the lunar month of Boedromion, were removed from underground pits and placed on the altars in order to be mixed with seeds. The underground pits served as the symbol of the Underworld where Persephone resides during the winter. The seeds, like Persephone, were kept in this sacred place to ensure a fertile crop in the spring. The priestess of Plouton, representative of the forces of the Underworld, conducted the ceremonies and the women mentioned above assisted her; unlike other Greek ceremonies, this was a festival of women (Dow and Healey 36).

In the "Hymn to Demeter," the ideology of the Thesmophoria is apparent. When Demeter discovers that her daughter has been abducted by the Lord of the Underworld, she wanders the earth disguised in the form of an old woman. She comes to the house of Metaneira in Eleusis and sits by "the Maiden Well" to grieve. Demeter, as crone, is approached by the daughters of Metaneira, maidens of the rarest beauty. Demeter is comforted by the maidens and is housed by their mother for a time. But when her stay is fruitless and the mortals cannot help her in her quest, she reveals herself as the goddess and curses them with winter retreating to her temple until Hermes comes to her with Persephone. Soon, Demeter forgives the mortals and teaches the families of Eleusis the mysteries of her rituals.

Demeter in the form of a crone plays a vital role in the mysteries. She is winter personified and the daughters who comfort her may very well be the symbols of the maidens who perform her ceremonies in the Thesmophoria. Again, their parallels in the sky are the Hyades. In a festival devoted to women on all levels, immortal, mortal and heavenly, the women are allowed to retain their Neolithic rights of the care and preservation of the grain. In "The Hymn to Demeter," Hecate, Demeter, and Persephone as the three aspects of the Pregnant Vegetation Goddess depict the feminine role in the grain mysteries. At the end of the myth, Demeter shares the mysteries of the grain with the families of Eleusis and the inherited Neolithic matriarchy is reinforced.

Despite the fact that we do not have the details of the Celtic rituals, the days on The Sequani Calendar are clearly marked as Tiocobrextio and three crosses signify the days when the full moon passes the Hyades in the month of Edrinios. Furthermore, the Hyades are marked as the guides for the lunar cycle. The concepts of the grain mother as crone and the giving of the gifts to the tribes are most evident in Celtic mythology as is the establishment of the goddess as sovereign of the land. The most ancient of mothers, the caillech, is the Celtic expression of the crone. She is found in numerous cites in Britain and the Continent.

In Ireland, she is known as the Caillech Beara and often referred to as the Old Woman or Hag of Beara. She originally appears as a tripartite goddess with the Caillech Bolus and the Caillech Corca Duibhne. In the book of Lecan, she is the mother of the tribes because she has seven periods to her youth, she marries seven husbands, and she has fifty foster children. The caillech, who presides in many local Celtic tribal icons, is often portrayed as the keeper of the woodlands and associated with sacred wells or river boundaries as a territorial goddess. According to Anne Ross, the caillechs, like the topographical goddesses such as Macha and Grian, became guardians of the wells or hill where they once propitiated (293).

The ancient goddesses of Celtic and Greek mythology retain the power of the mother goddess in her three roles as maiden, mother and crone. They are inherent in the belief systems of the ancient peoples because their stories are part of an agricultural cycle that is connected to the sun, the moon and the stars. In the lunar cycle of Edrinios, we have seen the last part of the agricultural myth re-lived in the heavens and on earth. The 55-night cycle is an event that marks the changing of the seasons. As a microcosm for the 55-year lunar cycle, it allows us to participate in a miracle of cosmic change. In this way, the cycle of the Fall Equinox becomes relevant on many planes of existence.

Chapter Two

Light Through the Hero's Eyes

In the month of Elembivios on The Sequani Calendar and the month of Boedromion on the Sacred Calendar of Eleusis, four constellations appear in the night sky that complement the picture of the myth of the ancient mother and daughter, Demeter and Persephone. These constellations are Cassiopeia, Andromeda, Perseus, and Cepheus. The myths of these constellations were written from the accounts of Herodotus, an historian of the fifth century BCE, where he establishes the lineage of Perseus. In the second and first centuries BCE, Apollodorus, Ovid, and Horace among others, told the tale of Perseus' birth and marriage to Andromeda. These myths may be seen as later Iron Age versions of the Bronze Age or earlier Iron Age myth of Demeter and Persephone. They are stories of the stars that emphasize the hero's journey and may therefore be imagined as light through the hero's eyes.

Cassiopeia is one of several constellations that are viewed continually as they circumnavigate Polaris, the North Star. Cassiopeia was said to have been "condemned by the gods to swing forever around the North Celestial Pole because of her offensive boasting" (Staal 14). At Elembivios and Boedromion, Cassiopeia is inverted and her stars form a "W," a symbolic posture to punish her for bragging. In different versions of the myth, Cassiopeia has boasted that she, and/or her daughter, Andromeda, are more beautiful than the Nereids, the naked nymphs who dwell at the bottom of the sea.

In both the "Hymn to Demeter" and story of Cassiopeia, the mother must suffer. The mother is punished for her beauty and the beauty of her daughter when the deities recognize her. In the "Hymn to Demeter," Hades abducts Persephone, and Demeter much search the earth in vain for her daughter. In the myth of Cassiopeia, Poseidon is angered by the boast that Cassiopeia's daughter, Andromeda, is more beautiful than the Nereids, his nymphs, and Cassiopeia must sacrifice her daughter to the sea monster, Cetus. The most significant difference in the myths is that the myth of Demeter and Persephone does not include the overstepping of boundaries by the mother, and the mother is a goddess, not a queen. Demeter's daughter is abducted by Hades with the permission of Zeus, almost as if it is part of a divine plan to find a wife for the Lord of the Underworld and exercise the sexual rites of the daughter.

Although these stories are descendents from earlier Bronze or Iron Age myths of the loss of the seed, or daughter, to winter, both myths retain the suffering and loss of the mother for her daughter as a reminder of the seasonal changes. The interest of the Underworld deity, Hades, leaves another basic element of the earlier myths intact: the necessity of the contact with winter and symbolic death represented as the Underworld. Loss and death dominate the myths with the arrival of the constellations of the Fall Equinox. The major difference is that the later myth assumes that the woman as queen, not goddess, is responsible for the suffering and is due punishment.

The daughter also seems to experience a rectitude of necessary quiet and suffering in the myths. In the myth of Persephone and Demeter, Persephone is presented as a sufferer who has taken her throne beside the god of the Underworld. In the "Hymn to Demeter," she is temporarily returned to Demeter as a vision of life-in-death. The sixth and seventh nights of the dark half of the lunar month of Boedromion are the nights in the Telesterion where the Hiera is shown to the people and the vision of Persephone is complete; the people know that the seed will be safely stored and kept for the winter months. In the later Iron Age myth of Cassiopeia, Andromeda is chained to a rock awaiting Cetus, a monster Poseidon created from his anger. In the sky, she is a long line of bright stars with fainter stars "that represent the outstretched arms of Andromeda while she is chained to the rock" (Staal 9).

Despite the common denominator of her suffering, the figure of the daughter in the myth of Cassiopeia and Andromeda is diminished from its original position in the earlier myth of Demeter and Persephone. In the "Hymn to Demeter," Persephone as Queen of the Underworld is transported by the Charioteer, Hermes, to her mother where they embrace in a touching scene of reunion. Demeter tells Persephone that Persephone must stay with her husband because she has eaten a pomegranate seed he gave her. She has a knowledge of daughter's her fate that has been planned by the gods and forces of creation, and she knows that Persephone has been given the title of Queen as compensation. Persephone and Demeter accept Persephone's fate, as we accept the oncoming winter months, and Persephone returns to be seated comfortably in the Underworld as goddess and queen.

In the later Iron Age myth, Andromeda is a woman and not a goddess, and her abduction is life-threatening. As she is to be sacrificed to the monster Cetus, she shrieks in agony for her parents, Cassiopeia and Cepheus, to save her from a violent death. Her position in the culture has diminished from a major goddess of the seasons of fall and winter, to a sacrificial virgin of the gods. Moreover, her political status has been diminished because she is the daughter of a foreigner, the King of Joppa in Ethiopia. In Ovid's translation of the myth much is made of Andromeda's suffering as a victim of circumstances, and in Apollodorus' version, Andromeda is seen as "prey" to the monster.

Although the representatives of Great Goddesses of the Grain Harvest, Demeter and Persephone, reenact their yearly cycle of separation and temporary loss for the benefit and knowledge of the time-keepers of the agricultural year, the ideology that remains in the mythology is obviously the suffering that the mother and daughter must face during their separation. In later versions of the same myth, the two women, Cassiopeia and Andromeda, must re-live the separation of mother and daughter. Where the earlier Bronze and Iron Age story must explain both mysteries, the life and growth of woman as earth and mother, the later Iron Age focuses on the human drama and its intensity. The focus on human drama is most naturally advanced with the appearance of yet another human who is connected to the gods: the hero.

The most significant addition to the myths and the constellations of the Fall Equinox is the appearance of the hero, Perseus. In the night sky, the constellation of Perseus "towers protectively over Andromeda" with one foot almost stepping on the star cluster of the Hyades (Staal 9). It seems as if the Hyades, who have been recognized in the rituals of the Thesmophoria in the first half of the month of Pyanopsia, are now dominated in this story in the sky. In his hand, is the head of Medusa, which includes Algol, a star that changes its brightness every three days as if its powers still resonate the awful power of the Medusa. Again, this magnificent appearance of Perseus addresses the fact that the full moon has passed the Hyades and the hero must cross the river Eridanus to begin his journey.

The addition of this figure to the pantheon of Fall Equinox constellations represents the addition of several concepts that permanently alter the path of the goddesses and gods of former eons. Primarily, a major figure of the Neolithic goddesses, as defined by Marija Gimbutas in *The Living Goddesses* as the Pregnant Vegetation or Earth Goddess, must give her powers to the hero in order that he may begin his journey. This is done through the Indo-European Ashvamedha-type ritual of the Sacred Marriage to the Horse Goddess and the subsequent transference of the power of that marriage to the hero.

The Earth Goddess and Goddess of the Grain Harvest, identified as Demeter Erinys in Greek mythology, gives her power as Horse Goddess in the Sacred Marriage, a position that is forced in much of the Indo-European additions to the myth, to the hero. The equine associations in the later Iron Age myths are transferred to the hero, and the hero uses the Horse Goddess' power to secure his mother and wife into happy marriages. Demeter's power as Earth and Grain Goddess must also be diminished to mortal stature in the figure of Cassiopeia; her daughter, unfortunately suffers the same fate.

The goddess Athena also comes to the forefront in the hero's journey where he harnesses her strengths to protect his people. Again, the Indo-European culture accepted the Neolithic Snake and Bird goddesses' powers of protection as sacred and found it necessary to transfer them to the hero. The image on the Aegis, the Medusa, and the protection of Athena are therefore central to the Perseus' success.

Perseus' power must then originate from the goddesses, often represented as women in his life, who bestow their gifts upon him. The first power bestowed is the transference of equine power. The transference of equine powers to the hero, Perseus, begins with several equine references in the myth of Perseus. When Perseus is born, a prophecy that he will murder his grandfather, Acrisius King of Argos, forces Acrisius to put both Perseus and his mother, Danae, in a chest and send them out to sea where they eventually float to the shores of the kingdom of Polydectes. When Polydectes tries to force marriage upon Danae and Perseus defends his mother, Polydectes pretends to desire another woman, Hippodameia, and asks each of his guests at an assembly to contribute one horse apiece as help for a love-gift.

Perseus cannot contribute a horse, and he desperately wants to relieve his mother from a forced marriage to Polydectes; therefore, his only option is to fulfill Polydectes' request to decapitate Medusa. The theme of the forced Sacred Marriage and the presence of horses as a means of power through dominance permeate the myth of Perseus from the beginning as if Perseus' fate is connected to the transference of equine power and the saving of the Earth Mother Goddess as his own mother. Hippodameia, the "tamer of horses," and Polydectecs are adversaries to his quest and yet they seem to aid him without knowing it. They send him on a journey that will more than likely be unsuccessful, yet it becomes his triumph. Instead of taming or dominating horses, Perseus gains their ultimate power and supernatural strength when he slays Medusa.

Medusa, an earlier version of Demeter Erinys, is said to have been raped by Poseidon who comes to her when she is in the form of a mare and he is a stallion. The products of this union are Pegasus, the winged horse, and Chrysaor, a warrior. When Medusa is slain by Perseus, Pegasus and Chrysaor spring forth from her neck. Pegasus is a boon to Perseus and a representative of the Earth Goddess' transference of her equine power. Moreover, Pegasus is a constellation that dominates the sky of the Fall Equinox. Pegasus is the front half of the horse which appears as if it is being born in the sky. The navel of the horse is the star of Alpheratz which is the crown in the constellation of Andromeda. The Greek astronomer Aratus calls Pegasus "the Sacred Horse" (Staal 29).

The birth of Divine Twins from the Sacred Marriage of the Horse Goddess may be traced to the earlier version of the forced marriage of Demeter Erinys and Poseidon Hippius where Demeter retreats to a cave near Phigalia and gives birth to the twins, Areion, a horse, and the Despoena, a girl with the title of "the mistress." The earlier version seems to establish the Despoena as a goddess in her own right with temples in Arcadia. Like Persephone, the Despoena and Andromeda are daughters of the Horse Goddess. In the constellations, the head of Andromeda comes from the horse's navel; birth ties the horse to the daughter as Divine Twins from the Sacred Marriage of the Horse Goddess.

According to Miriam Robbins Dexter in an article entitled "The Hippomorphic Goddess and Her Offspring," the myth of Demeter Erinys includes Demeter's retreat to a cave where she shut herself up. The Phigalians decided to carve an image of wood in the cave in the form of the goddess as horse with the head of a horse and the body of a woman; out of her head rise serpents and other animals (Dexter 287). The major difference in the earlier version of the myth and the Perseus myth is the absence of the hero. According to Dexter, the Indo-European proto-ritual includes the goddess' rape by a horse-god and the subsequent birth of her twins. The later myth involves a mortal heroine who "was slain by a hero, subsequently giving birth to a horse or being punished 'as' a horse. Only mortal heroines could reflect the true sacrifice of the mortal mare" (Dexter 294).

This important delineation in the evolution of the Sacred Marriage explains why Perseus must slay the mortal Medusa. His possession of her head as a weapon demonstrates the transfer of power to the hero. Perseus is now equipped with the powers of the Sacred Marriage. He has in his possession the head of Medusa and he owns the Sacred Horse. The cult of the head and the spiritual belief that the head contains the soul and ultimate power of the individual reinforces the idea that Perseus now holds the Medusa and the power to turn anyone who sees her to stone. In essence, Perseus temporarily holds the powers of life and death, powers usually reserved for the gods and goddesses.

With all his protection and powers, Perseus still has a bloody and hard-won battle to slay Medusa. Ovid gives the most poetic account of Perseus' battle with Medusa in Book IV of *Metamorphoses*, "The Story of Medusa's Head:"

> But in the mirror of his polish'd sheild
> Reflected saw Medusa slumbers take,
> And not one serpent by good chance awake.
> Then backward an unerring blow he sped,
> And from her body lop'd at once her head. (17-21)

The bloody battle ended, Perseus flies across the sea with Medusa'a head, an image immortalized in the night sky.

After slaying Medusa and completing a series of adventures, Perseus returns his powers to the goddesses. He returns the head of Medusa to Athena where she fixes it on her Aegis permanently as a symbol of strength, wisdom and protection. Athena, a representation of the Snake and Bird Goddesses of the Neolithic era, survived as a protection goddess in later eras. Here, she becomes a valiant icon in battle dress needed by the Iron Age heroes as a model for bravery in warfare. As protectorate of the city-states, Athena wears the Medusa as a protector of the people and the land.

Perseus also returns a shield of gold fashioned by Hephaestus that Athena gave him to confront the Graeae and the Stygian Nymphs who in turn lead him to Medusa. The Graeae, three swan-like sisters, fair of face but born old to only grow older, give Perseus directions to find the Stygian Nymphs who in turn relinquish their powers to Perseus in the form of gifts. The Nymphs give Perseus winged sandals, a cap to make him invisible, and a wallet or "Kibisis" to put the head of Medusa in (Apollodorus 2.4.2). Again, Perseus receives aid from goddesses and retains their powers. And finally, Hermes, or Auriga, offers Perseus help to fulfill his heroic destiny with the gift of a adamantine sickle, a symbol of the agricultural powers of Demeter.

With his acquired powers, Perseus is able to rescue Andromeda from the sea-monster, Cetus, and take her with him. In Ovid's *Metamorphoses*, much is made of the nuptials where Andromeda's former suitor challenges Perseus. Again, with the aid of Medusa's head, Perseus is able to turn his adversaries to stone only to return to Seriphos to battle Polydectes who faces the same fate. Perseus frees both his mother and wife from forced marriages and returns home to become the progenitor of a large family; Andromeda bears one daughter and five sons and they rule Argos for generations.

The Iron Age myths of the Celts express similar themes as those expressed in Greek mythology in terms of the hero's acquisition of the power of the Horse Goddess from the Sacred Marriage. Primarily, Celtic myths reflect the Indo-European ideology of the Sacred Marriage of the Horse Goddess to the King as a symbol of the goddess uniting with man to insure the wealth, success and continued fertility both for the king and the entire tribe. Irish mythology depicts this when Macha comes to Crunnchu to have her twins, Liath Macha and a son. In the Welsh *Mabinogion*, Rhiannon, in the form of a mysterious woman riding a white horse, marries Pwyll to give birth to Pryderi while Teirnon Twrvliant's horse foals a horse for Pryderi.

In both the later Greek and Celtic Iron Age myths, the Sacred Marriage to the Horse Goddess produces Divine Twins as a symbol of the powers of procreation and wealth and insures that this power be delivered to the hero in the form of a horse from the Twins. Where Perseus is given Pegasus, Cú Chulainn is given Liath Macha, or the Grey of Macha, and Pryderi is given the foal of Teirnon's horse. Unfortunately, the goddess must suffer in the myths in order for the heroes to assume her power. Like Demeter Erinys who suffers the loss of her daughter and Medusa who suffers the cruel fate of decapitation, the Celtic Horse Goddesses must suffer similar fates. Macha dies giving birth to her twins in the great race for King Conchobar, and Rhiannon suffers two-fold bearing guests on her back and being imprisoned and collared.

Macha's story is perhaps the oldest of the Celtic myths with equine references. Macha makes her first appearance in "The Coming of the Tuatha Dé Danaan" in *The Book of Conquests*. She is part of a triune of war goddesses and assumes her role as supporter and wife of Nuada, King of the Tuatha Dé Danaan. In the Iron Age myths of the *Táin Bó Cuailnge*, Macha's famous horse race takes place. In her most powerful role as Horse Goddess, Macha bears twins under much duress and dies in childbirth. Like Despoena, Demeter Erinys' daughter by Poseidon, Macha reveals that she too is the daughter of a sea deity, Sainrith mac Imbaith. Her final curse on the nine generations of Ulstermen remind them she is a powerful goddess who has suffered because of her marriage to a mortal.

Despite Macha's curses and undue suffering, she endows the hero with great wealth in the form of a horse. Liath Macha is given to Cú Chulainn who is free from Macha's curse, "the pangs of Ulster" (*Tain* 8). Liath Macha, like Pegasus, is one of the famous horses of mythology. Before Cú Chulainn goes into his final battle, Liath Macha sheds tears of blood for Cú Chulainn. At the battle, even when wounded, she is still able to kill fifty warriors with her feet and thirty more with her hooves before she dies at the Slieve Fuad (Ellis 118). She is a symbol of the Horse Goddess' power and is a constant companion to Cú Chulainn from birth to death.

Cú Chulainn's war-chariot, pulled by his two horses, Liath Macha and the Black of Sainglenn, is also a symbol of status for the hero. According to Dean A. Miller, the Indo-European chariot horses were used to transport the warrior to the battlefield (277). They were a symbol of power, wealth and strength in support of the hero as opposed to their later use in the calvary for actual fighting. The Celtic heroes would run and balance "the chariot's draught-pole at speed" as displays of the elite hero warrior (Miller 277). Likewise, in Celtic mythology the hero's horse is regarded highly almost as a companion and soul-mate with an attachment that resembles the attachment of twins. She is the hero's safeguard against all foes, a gift of the goddess.

In the Welsh *Mabinogion*, the Sacred Marriage of the Horse Goddess to the king, among other tales, depicts patterns of regeneration in seasonal cycles. According to Jeffrey, two men involved with one woman or goddess reveals a pattern of renewal with the changing of the seasons where the woman or goddess must be with one partner or the other depending on the season. Rhiannon, the Welsh Horse Goddess, is part of the pattern as well as Arawn's wife in the myth of "Pwyll Lord of Dyved." The changing of these patterns, remarks Gantz, is in yearly cycles where the heroes and goddesses travel to the Otherworld. Motifs for the seasons focus on the advent or absence of the goddess as a symbol of the regenerative force for change.

Gantz states: "This idea of a sympathetic relationship between the potency of the king and the fertility of the land is supported by Irish texts which hint that a king might be ritually married to the tutelary earth goddess of the tribe" (18). In the case of Rhiannon, the marriage takes place at the end of summer or early fall. Gantz parallels this event to the myth of Demeter and Persephone. More importantly, he adds that the names of Arawn and Havgan suggest fall and summer, respectively as Arawn appears in a grey-brown hunting outfit and Havgan's name might be translated as "summer white." Gantz believes that the Arawn-Havgan dispute is the basis for the changing of the seasons.

In the tale of "Pwyll Lord of Dyved," Pwyll encounters some Otherworld dogs, distinctively marked by their white coats and red ears, with their master from the Otherworld, Arawn of Annwvyn. Because Pwyll has hunted Arawn's stag, he owes Arawn a favor. Arawn asks Pwyll to take his place in the Otherworld to battle Havgan. Pwyll accepts Arawn's challenge and lives Arawn's life, except he does not have sex with Arawn's wife. Much like Persephone, Pwyll accepts his place in the Otherworld with its lush gold entrapments. However, Pwyll is a mortal and cannot stay in the Otherworld, therefore he does not taste the pomegranate or fruit that Persephone does; in other words, his refusal to engage in sexual union with Arawn's wife allows him to return.

Pwyll remains in the Otherworld for one year and one day, and then he must engage in battle with Havgan. In their combat, Pwyll mortally wounds Havgan with a single blow and although Havgan begs Pwyll to finish him off, Pwyll will not because he has vowed to Arawn not to do so. Pwyll's faithfulness to Arawn's request might be seen as an effort not to entirely destroy Havgan, as he is a representation of summer. Pwyll's fate is to merge the two kingdoms and not destroy the one. Pwyll and Arawn as winter figures must merge with summer to explain the seasonal changes that fall brings to the land of Annwvyn; "the name of Pwyll of Dyved fell into disuse, and he was called Pwyll Head of Annwvyn ever after" (*Mabinogion* 51).

The merging of Pwyll's kingdom and Havgan's kingdom may also be seen as the merging of the dark and light halves of the Celtic year or the waxing and waning of the solar deities becoming two halves of one circle. Pwyll's first success is followed with a second calendar event one year and one day later. Here, he is asked by Rhiannon to meet and marry her. Rhiannon appears to Pwyll from a mound, or opening of the Otherworld, to entice him into matrimony after a chase on horseback. Pwyll willingly agrees to not only meet and marry Rhiannon but to save her from a forced marriage to another figure from the Bright World or the world of light named Gwawl. The theme of the hero saving the Horse Goddess from marriage surfaces again as a dominant theme in both Greek and Celtic legend. Moreover, the darkness of winter must once more face the forces of summer white.

The marriage to Rhiannon is postponed until the third year in this calendar which corresponds with the celebrations of the Sacred Marriage to the Horse Goddess in the third year of the Sacred Calendar of Eleusis and The Sequani Calendar. Gwawl must be tricked into relinquishing his rites as both an adversary of marriage and a symbol of the summer before Rhiannon and Pwyll are married. At the wedding feast of Gwawl and Rhiannon, Rhiannon tricks Gwawl by asking him to fill a magic bag with food. The bag, much like Perseus' kibisis that holds the head of Medusa, is a gift from the goddess to ensure that the hero is successful. When Gwawl jumps in the bag on Rhiannon's command to press down the food that will not fill the bottomless bag, Pwyll storms the hall with ninety-nine horsemen and captures Gwawl. Again, at the request of the court, Pwyll does not destroy Gwawl in what the Welsh call the game of "Badger in the Bag."

In the third year of the marriage of Pwyll and Rhiannon, the suffering evident in all the myths of the Sacred Marriage of the Fall Equinox to the Horse Goddess is depicted. Rhiannon gives birth to a son and when the son is kidnapped, Rhiannon's maids, in fear for their own lives, smear the blood of animals on Rhiannon when she is sleeping and blame her for his death. Rhiannon's punishment is to "remain for seven years at the court of Arberth, where she was to sit every day by the mounting-block near the gate and tell her story to anyone who might not already know it; she was also to offer to carry guests and strangers to the court on her back" (*Mabinogion* 61). Although Rhiannon appears as a woman in the myth, it is here that her aspect of horse is apparent.

Her suffering ceases when Teirnon Twrvliant and his wife discover Rhiannon's son outside the stable of their mare who has just given birth to an amazing colt. The mare foaled every Belteine, a time for visits from the Otherworld, and usually, her colt was seized by a hand from the Otherworld. On this particular Belteine, Teirnon prevented the seizure and the boy was discovered. The myth concludes with Teirnon's recognition of the boy as Rhiannon's child and his subsequent return. Rhiannon is so pleased with her release from suffering that she calls her son, Pryderi or "anxiety," in memory of her trials. Pryderi rules the seven cantrevs of Dyved when Pwyll dies and becomes the hero that his father was.

The end of the myth of "Pwyll of Dyved," faithfully represents the birth of the Divine Twins of the Sacred Marriage. Pryderi and his colt are sent to the court of Dyved where Pwyll and Rhiannon welcome them into their lives. The colt is born simultaneously with the hero and acts as a trusted companion to the hero. Moreover, the powers of the Horse Goddess are successfully transferred to the heroes through the marriage of the King and his successors. Like Perseus and Cú Chulainn, Pwyll and Pryderi take their place as guardians of the Horse Goddess and her powers.

Through magic and myth, the story of the changing of the seasons is explained by the people of the Iron Age as a light through the hero's eyes. Although the myth varies from one culture to the next, the events are carefully marked on their calendars to ensure that future generations comprehend the cycles of time as they resonate in the sky. The traditional myths of the agricultural cycles depicted as the loss of the seed or daughter from the Great Goddess of the Grain Harvest are illuminated through the hero's journey and trials. The lunar months of Elembivios in The Sequani Calendar and Boedromion in the Sacred Calendar of Eleusis are demarcations of a grand event of celestial light and wisdom that comes to us through the stars reminding us that we are part of a cosmos much larger than ourselves.

CHAPTER THREE

The Great Goddess of the Winter Sky

A spectacular cycle of myths in the sky are the stories of creation at the Winter Solstice. They begin with the advent of Cantlos, the eleventh lunar cycle of The Sequani Calendar, and move through Samonios with the Solstice as its central focus, culminating with Dumannios and the advent of the cross-quarter days of Imbolc. These are the darkest nights of the year which are clearly marked by the passage of the moon through the winter constellations of the Pleiades, Orion, Castor and Pollux and finally, Orion's dogs, Canis Major and Canis Minor. More than any other season, the path of the moon illuminates the primary stars designated for each month on The Sequani Calendar by passing as close as possible to these stellar guides. The moon's journey therefore becomes the path of knowledge to the myths of this season.

The Great Goddess of the Winter Sky is the goddess of the moon known to many cultures by many names. In the Neolithic Culture described by Marija Gimbutas in *The Goddesses and Gods of Old Europe*, she is the Great Goddess of Life, Death and Regeneration who existed for at least five thousand years before the appearance of Classical Greek civilization in the fifth century BC (200). Her journey through the night sky of the winter months is a journey of the renewal of the life forces in their most dormant stage. She must therefore possess the capability to regenerate and renew life from bare sustenance or as we know it, from winter in its bleakest form. On the ancient calendar, the lunar cycles and stars that guide us invoke the image of the Great Goddess in the Winter Sky by depicting each of her aspects in the constellations mentioned above.

When the full moon passes between the Pleiades and the Hyades in the eleventh lunar cycle of the year, it is time for the celebration of the Great Goddess of the Winter Sky to begin. On The Sequani Calendar, this is the full moon of Cantlos, and on The Sacred Calendar of Eleusis it is the end of the agricultural calendar of the ancient Greeks containing the festival of the Skirophoria. It is a passage time from the season of the Grain Mother's harvest and the facing of the winter months, to the actual beginning of winter marked by the rising of the Pleiades in several ancient cultures. The full moon passes between the Hyades, the priestesses of the final harvest, and the Pleiades, the priestesses of the Great Goddess of the Winter Sky.

On The Sequani Calendar, the celebration lasts until the fifteenth of the month of Cantlos which is the end of the light half of the month or the half of the month of the waxing and full moon. In all five years of the calendar, it is marked as D Tiocobrextio, the culmination of a series of events marked by Tiocobrextio notations in a complex pattern throughout the harvest months of the calendar. One might suppose that these harvest festivals were times when the people gathered to celebrate the abundance of the tribes, the production of the harvest, and in the last phase, the storing and blessing of the seeds for the winter and the passage into winter itself.

On the Sacred Calendar of Eleusis, we are given vital clues to the ancient rituals of this passage time of the first full moon of winter. Here, the festival is called the Skirophoria in the month of Skirophorion. After a sacrifice to the Mother of the Grain Harvest, Demeter, a portion of her sacred gruel or mixture of several grains and honey is carried in skira to the temple of the goddess to be stored (Dow and Healey 39-43). A ceremony of priestesses dressed in robes of white, presumably priestesses of the Great Goddess of the Winter Sky, then perform a ritual to bless the fields. Cakes in the shape of serpents are baked in celebration of the oncoming winter season (Gimbutas 215).

Designated as a women's festival, the Skirophoria presents us with a vivid image of what might have been a celebration of the clear winter nights when the added luster of the stars enticed the women to welcome the Great Goddess of the Winter Sky. The priestesses themselves might be thought of as the physical representation of the Pleiades on the earth as they danced the winter stars into their world. The Pleiades are a cluster of dainty stars that glitter with great beauty (Staal 59). To astronomers, they are explosions of time looking into the creation of galaxies, a foreshadowing of the creation of light at the Winter Solstice that also represents the explosion of light of the original creation and the creation of the new year. The Pleiades light of the deep sky gives us a vision, not only of the season to come, but of the eternal restoration of the light of creation.

The heralding of the appearance of the Great Goddess of the Winter Sky is accompanied by the eating and/or disposal of sacred cakes shaped in the form of a serpents. This too is a vital part of the creation myths about to take place at the Winter Solstice. The serpent, a universal symbol of what Mircea Eliade calls the myth of the eternal return, disappears at this time of the year. Like the actual serpent who hibernates in the winter months, the serpent as a symbol of the constant force of creative energy, is also out of mind for the winter in these celebrations. In the Dogon culture of the Niger River Valley in northern Africa, the serpent is stored in the granary for the winter months to protect the grains. Ogotemmeli, a high priest of this surviving Neolithic culture, calls this the descent of the celestial granary. It too takes place on the eleventh full moon of the year (Griaule 45-6).

The primal fields thus being made ready for the cold winter nights, and the first rituals of the serpent contained in the granary commenced, the celebration of the Great Goddess continues in the dark half or the waning half of the month with the setting of the Summer Constellations and the rising of the Winter Constellations. This major shift in the heavens is accompanied by the magnificent constellation of Orion rising in the night sky. Orion, one of the most spectacular and easily identifiable constellations is one of four constellations, Orion, Castor and Pollux, Canis Major and Canis Minor, appearing together in a pattern intense and wide that paints a portrait of the Great Goddess of the Winter Sky. So powerful is this image that many ancient monuments, including the Egyptian pyramids on the Giza Plateau and the granaries of the Dogon, are aligned to Orion's stars (Cornelius and Devereux 137; Griaule 32). Like the ancient monuments of the Celts, such as Newgrange and Stonehenge, they mark the Winter Solstice.

The Winter Solstice to the ancients was both a celebration of the bringing in of the new light of the year, as it is the time when the light returns and the days grow longer, and the celebration of the shifting of the constellations from the Summer to the Winter Constellations. The mounds, circles and pyramids of the earth are made to reflect this time when the earth and the sky meet to form creation. The best representation of this celestial event in the culture of the Neolithic peoples is embodied in the Great Goddess of Life, Death and Regeneration. According to Gimbutas, the Great Goddess of Life, Death and Regeneration is "a composite image with traits accumulated from both the pre-agricultural and agricultural eras" (152). As Lady of the Beasts, she represents the pre-agricultural goddess of wild nature and as agricultural goddess, she represents regeneration, fertility, and both the dark and light aspects of the moon.

Gimbutas describes her shape as "corpulent" with a massive structure and a phallus-shaped head inferring her androgynous nature (152). It is not too far-fetched to compare her very shape to that of the constellation of Orion which was only named that much later in the Iron Age. The two stars on The Sequani Calendar that mark the constellation's arrival in the winter sky are the broad shoulders of the Neolithic goddess. Betelgeuse and Bellatrix, whose name is "the female warrior," are these two stars. Furthermore, the belt of Orion, made of three very bright stars, accentuates the sacred three of the goddess and her wide hips are easily identifiable in the sky.

The myths surrounding the constellation Orion in both ancient Egyptian and Greek culture attest further possible connections with the Neolithic goddess. In Egyptian culture when the body of the pharaoh was placed in the King's Chamber, his phallus was oriented along the belt-stars of Orion and as the light for these stars traveled down the shaft of the pyramid, the pharaoh inseminated the constellation insuring his own resurrection (Cornelius and Devereux 77). Like the Great Goddess of Life, Death and Regeneration, Orion is responsible for resurrection in Egyptian mythology.

The image of Orion in Greek mythology as The Mighty Hunter reinforces that aspect of the goddess as Lady of the Beasts. With her two dogs, Canis Major and Canis Minor, accompanying her, the goddess embodies what Gimbutas sees in the ancient art of the Neolithic peoples as the force of regenerated life through physical strength represented by her two beasts. The myths of Orion affirm his connection to this ancient image by having his protection goddess as Artemis, the goddess that Gimbutas attests is the later Greek version of the Great Goddess of Life, Death and Regeneration. Although the passage of time and the changing images of the goddess in different cultures reflect the diversity of the human imagination, her essential energy and form is one compatible with the winter sky.

According to Gimbutas, several aspects of the Great Goddess survive in later mythologies. Three of the most prominent of these being her Mother Terrible aspect entwined with her Regeneration aspect or what Gimbutas identifies as a Hecate-Artemis nature; her birth-giving aspect or splitting of the egg into halves or twins; and her bear or bear-nurse aspect which serves to protect the newborn of the year (196-200). Each of these three aspects have parallels in celestial events. The Hecate-Artemis aspect is seen as the dark and light halves of the cycle of the moon accented in the winter sky by the moon's journey through the winter constellations of Orion and his dogs, Canis Major and Canis Minor. The splitting of the egg is the division of the year at the Winter Solstice represented by the constellation of Castor and Pollux, the Divine Twins of Greek mythology, and the nurse aspect represented as a bear, is celebrated at the cross-quarter day of Imbolc, a demarcation of the mid-point of the Winter Solstice and the Vernal Equinox.

On The Sequani Calendar, these celestial events are marked in the first two lunar cycles of the year, Samonios and Dumannios. On a calendar as sophisticated as The Sequani Calendar, the path of the moon, the constellations and the solar events are clearly marked for each aspect of the goddess. Moreover, Intercalary Time is celebrated every five years to reflect on the regeneration of cosmic time and original creation in which the goddess ultimately is the center. The Intercalary Time is a fifth dimension to the four seasons and an important aspect of the calendar. Pragmatically, the thirteenth lunar cycle that Intercalary Time is composed of is celebrated at the solstices in order to make the solar year coincide with the lunar. This must be done by marking Intercalary Time and "saving" it for the solstices approximately every two and one half years to create a five-year calendar that uses the lunar cycles as months of the year.

Intercalary Time is sacred time at the Winter Solstice. It represents the original creation of the universe as well as the creation of the year. The Creation Myths of the ancient peoples are reserved for this time. It is a time to worship the Great Goddess as the eternal return of the year and the cosmic return of the energy of the life forces represented by the celestial bodies.

In Vedic cosmogony "Prajapati" is the fire-altar of the New Year with its five cardinal points of the five directions. Four seasons or cardinal points and a fifth upper region are the basis for the altar to the gods and the belief in the eternal return of life through the year and throughout time (Eliade 79); Prajapati is the constellation of Orion (Chartrand 562). Likewise, at Stonehenge there are five Bluestone Trilithons at its central altar. In a sense, the year spirals towards the heavens and infinity in Intercalary Time.

Together with the celestial bodies, the picture of eternal creation with the goddess at its center explain the myths of the splitting of the year. The first event is marked on The Sequani Calendar on the third night of Samonios as EXINGI, the time to extinguish the fires of the old year. On the second night of the dark half of the month of Samonios, TRINOVX SAMO or PRINO SAMO, presumably are the markings to rekindle the fires of the New Year. The thirteen nights in the interim between these two holidays on the calendar have parallels in many ancient cultures and are known as the Twelve Days of Christmas in Christianity. In the tribe of the Tewa of New Mexico, they are called The Days of the Sun and in the calendar of the ancient Babylonians, they are known as the twelve days of the Akitu or the ceremonial epic of creation (Suzuki and Knudtson 59; Eliade 55).

For the Celts these nights are essential to their calendar. At Newgrange, the light of the first rays that rekindle the year pass over a line demarcated on a balain at the entrance of the passage tomb. The light splits this symbolic balain which is aptly decorated with the spirals of time marking the beginning of the light half of the year. Then, the light pierces the darkness of the passage mound through a roof-box to illuminate the back chamber where a womb-like seat of stone sits, an accurate metaphor for the seed inside the mound of the goddess waiting to be inseminated with the light of renewal and rebirth. This fantastic event for the ancients is a symbol of the rekindling of the light of creation, fire and energy of the solstice (*Archaeology Ireland* 15).

Newgrange, a magnificent passage tomb and one of twenty in Ireland, was built by the Neolithic peoples to presumably honor the Great Goddess of Life, Death and Regeneration. It most accurately represents her as the Bringer of Light to the new year through its symbolic representation of the cosmic and eternal spiraling and return of the sacred light of the Winter Solstice. To the Neolithic peoples, the Goddess as Bringer of Light is the magician-mother who restores new life to the dead. The Neolithic peoples buried their dead in a position similar to the icons of the goddess "squeezed into egg-shaped pithoi for burial" as if they were placed in the goddess' womb for rebirth (Gimbutas 159). Like the womb-shaped stone seat in the cruciform chamber of Newgrange, these ancient burials reflect a belief in the birth-giving or regenerative aspect of the Great Goddess.

"Because her main function was to regenerate the life forces, the goddess was flanked by male animals noted for their physical strength" (Gimbutas 195). In the sky, the regenerative forces of strength are the dogs of Orion, Canis Major and Canis Minor. They are marked on The

Sequani Calendar at TRINOVX SAMO, the beginning of the dark half of the lunar cycle of Samonios, by two stars, Sirius of Canis Major and Procyon of Canis Minor. As if to dominate the dark half of the moon, these stars protect and illuminate the winter sky at that time. Together with Orion, they are the largest collection of the brightest stars in any of the skies; they are a huge configuration that straddles the Milky Way. The Dogon of Northern Africa mark the cycles of the dog star, Sirius in terms of the larger cycles of cosmic time which corresponds to its fifty year orbit with its companion star (Cornelius and Devereux 79). The dog, therefore has both stellar and cosmic connotations to ancient peoples as a time-keeper of both the Winter Solstice and a time-keeper of the larger regenerative cycles of creation.

According to Gimbutas, "The dog, the howler by night, was the goddess' principal animal" (169). The dog belonged to the Moon Goddess and was her sacrificial animal. It was a protector of the fields and of any evil meant to man because it represented the forces of the nocturnal realms. Gimbutas believes that in her most terrible aspect the goddess must have been a "Mother Terrible yearning for human and animal blood, as indicated by her epiphany in the shape of a ferocious dog" (196). In later mythologies, Gimbutas traces the evolution of this regeneration aspect of the Neolithic Great Goddess that contains both dark and light phases to the Minoan and Mycenaean images of her as a Hecate-Artemis figure in their mythologies.

In Minoan Crete, the Great Goddess is represented as a Mistress of Animals on many of the frescos, rings and seals. She is often depicted with animals on either side of her or in ritualistic scenes of dance or in the hunt. She is often accompanied by her nymphs or priestesses, and her sacred places are the tops of the mountains. She is responsible for the life and death of humans and animals and is represented as the light and dark aspects of the moon. In some cases, she stands in a chariot drawn by stags or panthers, and in others, she holds a hind or small wild animal as if claiming dominion over it. Her name appears as A-ti-mi-te and A-ti-mi-to on Linear B tablets from Pylos (Gimbutas 197).

According to Gimbutas: "The image of the Hekate-Artemis of Caria, Lydia, and Greece, based on descriptions of early Greek authors, vase paintings, and finds in actual sanctuaries dedicated to this multifunctional goddess, supplement and verify our understanding of the appearance and functions of the prehistoric goddess" (197). Whether Hecate or Artemis appears as one or two goddesses, they are both connected with the two faces of the moon. The life-giving aspect is Artemis as the full moon and the death aspect is Hecate as the new moon. The constellation of Artemis is Orion and the constellations of Hecate are Canis Major and Canis Minor.

Gimbutas notes that in Western Turkey, Hecate was the primary goddess in the sanctuary at Lagina where dogs were sacrificed in honor of her. Hecate could turn herself into a dog. Likewise at the Zerynthos cave on the island of Samothrace, dogs were sacrificed and dances were performed; "Hecate and her dogs are described as journeying over the graves of the dead and above the sacrificed blood" (Gimbutas 198). It is almost as if they fly through the winter sky in the dark night above us, protectors of the dead with Sirius, eye of the dog, watching over the ever-changing cycles of human and cosmic time.

Whether negative or positive, dark or light lunar energy, both Hecate and Artemis are linked by their association as the carriers of light through the darkness. Hecate brandishes a torch as the mistress of the night roads and as the guardian of the crossroads of fate in the world of the dead. Hecate's torches were carried over the fields in rituals to bless the newly sown ground (Gimbutas 198). Martin P. Nilsson calls them the Aiora Festival where virgins, or priestesses, walked singing around the vine-hills and sacrificed a dog to promote the harvest (64). As regenerator of light in darkness or life in the fallow winter fields, both Artemis, The Bringer of Light, and Hecate, the torch-bearer and Queen of the Ghosts, our guide to the ethereal regions of the dead, are essential figures in the theogony of the Great Goddess.

In Mycenaean and Hellenistic Greece, Artemis and Hecate evolve into separate goddesses; Artemis, still predominately referred to as The Bringer of Light, is a twin to her brother Apollo. Artemis is given "the office of bringing light" by her father Zeus, and she is given a silver bow, ten hounds, sixty ocean and river nymphs as her attendants, and all the mountains in the world to dwell upon. In her golden chariot drawn by two hinds, she cuts her first torch on Mysian Olympus and avoids men by disguising herself and her attendants with faces of white mud or avenging herself by hunting them down and ripping them to pieces (Graves, Vol. I, 83-5). Now a daughter of the gods, and a sister to Apollo, Artemis tries to fiercely maintain her virgin aspect through her hunting and athletic abilities. In this demoted position, she is still able to control her domain of the Mistress of Animals through strength and independence.

Hecate, on the other hand, is able to maintain more of her earlier stature than Artemis. In the creation myths of the Hellenic Greeks, Hecate is a Titaness, one of the ancient Greek deities who preceded the Olympians. Although the other Titans are punished by Zeus and the

Olympians, Hecate retains much of her honor and is not violated sexually or robbed of her former powers. According to Miriam Robbins Dexter, "the potency of Hecate was reflected in her transformation into a transfunctional goddess" (124). She is pictured seated in judgement at the Assembly of gods and goddesses and respected for her wisdom. Furthermore, she fulfilled "a martial function" as the power behind the warriors who bestows victory and grants glory. She has the power as a goddess of regeneration to give and take life and because of this she is transformed as a goddess of the Netherworld which, according to Dexter, is not uncommon for a goddess who reigned over heaven, earth, and sea (126).

As goddess of the Netherworld and a representative of the final phases of the dark half of the moon, Hecate becomes a crone figure and her descendents become monsters in the later myths of the Greeks. Although Hecate maintains her affiliation with travelers and crossroads, she is feared as well as honored. Her emblem in the sky, Canis Minor, becomes Scylla, a once beautiful daughter of Hecate turned into a dog-monster with six heads and twelve feet. Changed by the jealous sea-goddess, Amphitrite, she would seize sailors, break their bones and devour them (Graves, Vol. II, 262). Her children, the Empusae, are also monsters and seducers of travelers; they are disguised as bitches, or beautiful maidens (Graves 263). Even her dog, Cerberus, a three-headed monster that guards the gates of Hades, is described as a fifty-headed monster retaining the fifty as a symbol of the fifty year cycle of Sirius.

The island of Samothrace, an island off the mainland of Thrace, points to interesting speculation. Hecate's cave, the Zerynthos, is the derivative of her name, Zerynthia, and a major site for her mysteries. According to Dexter, names like Zerynthos that end with "nthos" indicate that they are a part of a pre-Greek, perhaps Old European language, establishing Hecate as a goddess of pre-Greek origin. In order to incorporate the pre-Indo-European image of the goddess into Hellenic mythology, Hecate's island is revisited by the Argonauts. When Castor and Pollux arrive with Jason and the Argonauts at Samothrace, they are introduced to the mysteries of the cult of Hecate and each of the Dioscuri is given a star as his emblem in the sky. In the month of Samonios on The Sequani Calendar, the Twins are the guiding star for the month, and as stated above, Hecate is the guide for the dark moon of the month. The name of the island might have been derived from the month's root "samo" and the geographical proximity of Thrace thus re-uniting this ancient goddess and her island with the moon and the stars in the celestial year.

Orion makes his appearance in the mythology of the Iron Age Greeks as the hero associated with celestial light. Like Artemis, Orion is a great hunter and when asked to free the land of Chios of wild beasts in order to marry Merope, Orion almost completes his task. Oenopion, Merope's father, gives Orion wine and Orion violates Merope in his drunken state. Oenopion punishes Orion by blinding him and Orion must travel to the east to meet Helios where he first rises from the Ocean, to regain his sight. Orion meets Helios, has his sight restored, and falls in love with Helios' sister, Eos, the Dawn (Graves, Vol. I, 151). In the beginning of his mythology, Orion is established as a celestial deity associated with the sun, the dawn, and the east. His restored sight is symbolic of his resurrection and power attained from these celestial deities. Furthermore, his associations with the dawn and the east are symbolic of the bringing of the light of the new day, the new year, and the eternal return of the light of time.

Having thus attained the celestial powers equivalent to those of the Great Goddess herself, Orion is able to seek revenge on Oenopion and chases him to Crete where he meets Artemis and falls in love with her. Almost her equal, Orion, "the dweller on the mountain," shares the love of the chase and the hunt with Artemis. More of a son or twin than a lover, the two are symbiotic immediately. Forgetting both Merope and Eos, Orion goes with Artemis to hunt and sport together. Unfortunately, their love is never consummated because Apollo and Zeus fear the union and send a scorpion to kill Orion. Artemis pleads with her brother and father to immortalize Orion as a constellation where the scorpion, known as the constellation of Scorpio, still pursues him in the sky. Orion in his eternal pose as the hunter and the hunted is empowered and immortalized by the celestial deities of the light.

With a masculine figure replacing the Great Goddess of the Winter Sky, the Greeks introduce a myth to protect the Pleiades, the priestesses of the goddess and representations of her virgin aspect, from his amorous advances. The last myth of Orion involves Orion chasing the Pleiades, daughters of Atlas and Pleione. One of the Pleiades is Merope, the virgin that Orion violated and could never marry; she is called "the lost Pleiade" because she is a variable star. Orion is said to have chased the Pleiades across the meadows of Boeotia, until Zeus took pity on them and changed them into stars. Orion's vain pursuit of the Pleiades refers to the fact that they rise first in the winter sky and he rises in the east directly after them. The Greek myth accents the fact that the priestesses are now under the protection of the patriarchy.

The image of the Great Goddess of the Winter Sky is charted on The Sequani Calendar and represented throughout the mythology of the Celts. The Neolithic sites of Newgrange and Stonehenge as well as many of the stone circles and mounds throughout Celtic Europe and the British Isles reflect the immense importance that the people placed on the deities of the Winter Solstice by their frequency and size alone. Moreover, they were great gathering places where the stones marked the ritual observances of the Holy Nights of the Solstice. In the mythology, the same reverence for the goddesses and gods of the solstice is seen. The first aspect of the Great Goddess of the Winter Sky may also be traced in Celtic mythology as The Bringer of Light and celestial guardian of the new light of the eternal return.

At Newgrange, presumably named after her, she is known as Grian, the goddess of the light or sun as the Goidelic word "grian" means sun. The fact that the Iron Age Celts took the name of one of their most important monuments from its original Neolithic goddess establishes her importance to the peoples who originally built this great stone mound and her importance to the ancients who still used it. Later in the Iron Age mythology, Grianne is the daughter of the High King of Ireland, Cormac Mac Art and the wife of Fionn Mac Cumhail, the great hero of the Fianna warriors. In Celtic culture, the sun-room or gathering place for women to sit and sew or talk honors the name of the goddess and heroine by retaining the name of the Grianne. At Tara, there are three mounds called the Grianne, also in honor of the goddess.

In *Pagan Celtic Britain*, Anne Ross attests to the fact that there is very little evidence other than the actual topographical legends and place names for the goddesses. Even iconographic evidence is essentially limited. However, Ross believes that this is compensated for by "the wealth and apparent genuiness of the Celtic goddesses as they appear in the vernacular literatures of Britain, especially in those of Ireland" (292). The "genuine memories" contained in the place names and folklore point to the fact that goddesses such as Aine, Grian, and Macha found a permanent place in the mythology of these ancient peoples who associated them with the mounds and other natural and man-made sites (293). This is especially true in the case of the goddesses who protected and nutured the land.

In the mythology of the Celts, the deities associated with the sun are Grannos, Grianainech or Ogma, Belenos, and Lugh, all masculine figures. Grannos, the Gaulish sun god and god of healing is paired with

Sirona, whose name means star. His association with a celestial deity affirms his possible connections to the Great Goddess of the Winter Sky. A ritual chant to Grannos: "Granno, my friend, Granno, my father, Granno, my mother" also gives evidence to his original associations with the goddess (Ellis 118). Like Lugh and Apollo, Grannos is healer as well as a sun deity. Perhaps the healing attribute of these gods represents the regeneration aspect of the goddess.

The three remaining deities, Grianainech, Belenos, and Lugh are solar deities; however, none are particularly associated with the winter goddess. Grianainech "of the sunny countenance" is the most interesting in terms of the calendar because he is associated with Hercules. Like Hercules, Grianaicech's role as solar deity in the age of heroes might have been to trace the pattern of the hero's journey through the year. In Ireland, he is known as Ogma, the inventor of Ogham script, and in Britain, he is known as Ogmia, a deity of the Sol Invicta type. Belenos and Lugh are also associated with solar holidays, the cross-quarter day of Belteine and the cross-quarter day of Lughnasad, respectively.

If the light aspect is represented in Celtic culture by Grian, then the dark aspect or the Hecate of the Hecate-Artemis aspect, is represented by the dog in ancient Celtic culture. At the spring of Sequana, many ex-votos were found of people holding a dog (Ross 424). These ex-votos are similar to the Great Goddess figurines as the Mistress of Animals found in Minoan Crete and Anatolia that hold a small wild beast as if to have dominion over it (Gimbutas 156; Nilsson 505). Like the animals that flank the goddess, the dog is protector and companion of the goddess, a representative of her association with the beasts. In the mythology of the Celts, Cú Chulainn's appellation is "Hound of Culann" because, after killing Culann's hound, he becomes the protector of Culann's property, and Fionn Mac Cumhail's two companions are the dogs: Bran and Sceolang, his nephews.

The iconography of the dog also places it at scenes of death, sacrifice and the underworld. On the Gundestrap cauldron, the dog is present at two sacrificial scenes and in Britain at the temple of Nodons, a bronze dog was found at the bottom of a funnel in the center of the temple. According to Ross, "this suggests that this was an offering to the deity, conceived of as being situated down through the earth, much as the deities of springs and wells were supposedly lords of the otherworld below the ground" (424). These Celtic dogs, like Hecate, are representative of the powers of death and the underworld. They represent the death and darkness of winter and the control the goddess holds over this domain. The dog at the center of the funnel recalls the snake buried at the center of the granary as guardian of the chthonic realms.

The second aspect of The Great Goddess of the Winter Sky is marked on The Sequani Calendar by the rising of Castor and Pollux; Pollux is the star of primary magnitude or guiding star of the month of Samonios. These stars rise in the east on the first day of the month and guide us through the lunar cycle. They represent a second aspect of The Great Goddess which is her birth-giving aspect or splitting of the egg into halves or magical twins. The cosmogonical idea of the splitting of the primordial egg finds its roots in the Paleolithic Era where images of water, a water bird, eggs, a doe, and a woman are found. In the Neolithic Era, vase paintings and frescos also reveal the formation of the world from the primordial egg splitting in two (Gimbutas 101-102).

In Neolithic art, the beginning of life is depicted as an egg enveloped in water with a snake, the symbol of life energy, winding around the cosmic egg: "The Lepenski Vir sandstone sculptures dating from around 6000 BC are almost all egg-shaped and one has a vulva engraved in the centre" (Gimbutas 102). Likewise, the creation myths of most ancient mythologies include similar images of a cosmic egg splitting and a snake as the life energy force. A water bird who lays the cosmic egg is a vital part of the myth. She is often depicted with the egg inside of her body; in a similar manner the goddess depicted with enlarged buttocks is an abstract of this bird goddess with her egg (Gimbutas 107).

On the figurines of the Great Goddess of Life, Death and Regeneration two large egg circles represent the goddess' potential for birth as do her egg shaped thighs and buttocks. The double egg often appears with the goddess' other companions: her dogs, deer, and the moon. If the figurines are cut open, Gimbutas remarks that one actually finds a double-egg inside (166). These "exquisite designs" reappear in Minoan-Mycenaean art as well where their association with the goddess' aspect of regeneration is clear: "The constant portrayal of twin crescents, or two does with opposed bodies reflects the magical potency of a splitting pair" (Gimbutas 196). As container of the splitting egg, the goddess herself is the creative principle. Unlike her counterparts in the later mythologies of the Greeks and Celts, she does not need a male sky-god to become pregnant and give birth to the cosmos.

The birth of the twins is clearly depicted in the night sky of Samonios. Here, Castor and Pollux, representatives of the splitting of the egg, are twins linked by their hands. They emerge from the Milky Way, the Great River in the Sky. Where the Milky Way splits, like a great river splits at a ford, the constellation of Cygnus, the Swan, sets in the west. Cygnus is flying below the horizon and will not rise again until spring where she is marked on The Sequani Calendar as a primary guide in the spring and summer months in the forming of the Summer Triangle of the Bird and Snake Goddess. Like actual water birds, such as the swans and geese, she has migrated to a landing to lay her eggs. This portrait in the sky tells the story of the birth of the twins from the splitting of the cosmic egg of the water-bird goddess.

In the mythology of the Indo-Europeans, this story undergoes many adaptations. Instead of the image of the twins being born from the egg of the water-bird, the twins drive a solar chariot that chases the swan or is drawn by swans. The addition of the solar chariot obviously reflects the Indo-European use of the chariot, but the addition of the solar associations effect the cosmology. The father god of the diurnal sky, **deywos*, is the husband of the year, **yera*, and he must marry his wife each year: "The meaning of this legend is beyond doubt the yearly reappearing of light after the winter night" (Haudry 229). In Greek mythology, this is the marriage of Zeus and Hera, Hera being a representative of the Bird and Snake Goddess of the Neolithic pantheon. Instead of the Great Goddess of the Winter Sky and the Bird Goddess of the Summer Triangle uniting to form the Winter and Summer Months of the Year of Creation, the sky father and his twin sons are conquering the year in their solar chariot.

The heroes of year, such as Hercules, foster-son of Hera, must also conquer the year by escaping their death and surviving until the Summer Solstice of the second year or to the Intercalary Time symbolic of immortality. To do this, they must swim the winter night: "An Indo-European kenning may be reconstructed as 'to swim through the winter night' " or "aptoryama, to ford through the water" and attain the safety of the shore of the year (Haudry 230). The Milky Way, symbol of the River in the Sky, depicts such a ford where the Swan clears her way at its dividing point in the night sky of Samonios. In the Indo-European texts, the heroes must pass through the same stream to achieve immortality.

While the heroes are busy swimming the river like the ancient goddess, the Rbhavah, who are the Vedic gods of the year, are sleeping twelve days in the house of Agohya, the one who should be concealed or the sun. Like the thirteen nights marked on The Sequani Calendar representing the nights where the fires of the year are extinguished and re-lit in honor of the Great Goddess of the Winter Sky, the Indo-European tradition calls for the same observance by their gods. This sleep of the year, or thirteen-night vigil, takes place at the end of the world or below the horizon, the place that Cygnus flies. After these twelve days, the Vedic gods swim through the winter night to reach the light of the Diurnal Sky.

The apparent changes from a lunar and stellar calendar that measures the night sky by its constellations of the goddess to a solar chariot and gods and heroes who command the same rituals under masculine guise and solar holidays, changes the calendar and mythology of the ancients significantly. The myth of the Divine Swan and the Twins, however, remains in the night sky forever and our understanding of the constellations and the lunar cycles can easily be retrieved through the knowledge of astronomy and mythology on The Sequani Calendar. Furthermore, the evidence of the Divine Swan in the myths of the Twins can be traced through the mythology of the Greeks and the Celts in its evolved form in the written texts.

In the myths of the Divine Swan and the Twins, the water-bird becomes the star sister of the twins, a minor deity, not the Great Goddess who is able to give birth parthenogenetically. The two young men, regarded as offspring of the solar deity, "share a consistent relationship with the Sun god and the goddess or goddesses who represent the dawn, the morning and evening stars." These goddesses are represented as their sister, consort, or wife and her "epiphany" is frequently a swan; she is identified with words that mean "bright," "white," or "shining." In the myths, the twins must often rescue their consort from some sort of watery peril by traversing the sky in the day and resting with her at night (O'Brien 161).

In Greek mythology, the tale of the water-bird goddess and her twins has its roots in their creation myths. In The Pelasgian Creation Myth, Eurynome, the Goddess of All Things, couples with the great serpent, Ophion and assumes the form of a dove who lays the Universal Egg that splits in two. In this myth, the dove is a water bird who lays the egg on the waves. In the Homeric and Orphic Creation Myths, the goddess Night was courted by the Wind and laid a silver egg in the Womb of Darkness; eventually her sceptre is passed to Uranus and he rules the world in her stead. Uranus, the sky father, marries Rhea, mother of Hera, just as Zeus marries Hera (Graves, Vol. I, 29-32).

These myths retain the image of the goddess as water-bird giving birth parthenogenetically, until the change in calendars where the solar fathers and sky deities change the emphasis of the creation. Zeus couples with more than one-year goddess in a myth that recounts his adulterous relations. He changes into a swan to court Nemesis, the goddess who wears a silver crown adorned with stags, and carries the Wheel, a symbol of the year; Nemesis has shape-changed into a goose, so Zeus must also change into a bird to overcome her. Leda, a Spartan woman, finds the hyacinth-colored egg that Nemesis lays, and she brings it home and hides it in a chest. In other accounts of the myth, Zeus in the form of a swan violates Leda herself. From the coupling of Zeus and Nemesis or Leda, the twins, Castor and Pollux, and Helen are born (Graves, Vol. I, 206).

The story of Castor and Pollux and their sister Helen, a later variant of the myth of the Divine Swan and the Twins, also has its roots in earlier Greek mythology. In the myth of Eos, the Goddess of the Dawn, Eos meets her brother, Helios, the sun god, and they travel in their chariot each day to bring in the light and accompany it across the sky. At night, they sail home along the Ocean Stream, the Milky Way, after resting in the Islands of the Blessed, the islands representing eternity that the heroes swim through in their journey of the winter night (Graves, Vol. I, 149). Their daily journey is symbolic of bringing in the light of the day, the light of the year and the light of the eternal return; however, the myth no longer emphasizes the goddess as its center.

In the myth of the birth of the twins, Castor and Pollux, their sister, Helen, whose names means "light" and is etymologically related to the name of the sun-god Helios, is hatched from the swan and goose's egg to later be rescued by two other brothers, Menelaus and Agamemnon, in the *Iliad*. As an Iron Age version of the bringer of light, she is shadowed by her crime of adultery, and she is reduced to mortal status even though she is still the most beautiful woman on earth. Like the rivalry between Castor and Pollux and Ida and Lynceus, the Iron Age myths of the twins and their sister include clan warfare as their central theme. At their death, their image is set in the winter sky as a reward for their devotion to brotherly love and clan unity.

In Celtic mythology, the myth of the Divine Swan and the Twins begins at the Winter Solstice at Newgrange. It is marked by the monument and by the migration of the swans to the fields around Newgrange where they nest for the winter months. In fact, Newgrange is probably the only site in County Meath where Whooper Swans come on a regular basis each winter and in recent years number as many as 226 in their flock (Murphy and Moore 2). The actual swans, the swan in the sky, and the celestial light splitting the balain at Newgrange combine to form an everlasting image of the Winter Solstice. When the constellation of Cygnus is setting in the west, the myths of the swan and the solstice begin.

According to Anne Ross, the earliest evidence of a solar cult associated with the swan is found in the votive models of carriages drawn by stags, oxen, or swans from the Urnfield and Hallstatt phases of European pre-history. The swan or swans "figures in conjunction with the carriage which it either pulls or on which it perches." These sacred swans are often linked together by chains or have neck-rings as decorations. A horned beast, who is also chain-bearing, accompanies the chariot which is often decorated with solar or astral symbols (310). This composite image of cult animals closely parallels the images of the Great Goddess of the Winter Sky from the Neolithic Era. The chariot is the goddess' vehicle to bring in the light of the solstice, the animals that flank the goddess are the beasts of the goddess, and the paired birds are the swan and twins.

On several of the votives, the astral chariot is driven by a figure in the form of a beaked male or a divine priest-king. Ross believes that this is also evidenced in the late Iron Age mythology of the Celts where the chariot is driven by the hero, Cú Chulainn in the *Táin Bó Cuailgne.* Here, the same composite cult image is presented when Cú Chulainn masters the wild beasts returning to Emain Macha in his chariot. Cú Chulainn has fastened two wild deer, twenty-four wild swans, and his horses as well as the three heads of the sons of Nechtan Scene to his chariot. When the beasts become unruly, Cú Chulainn tames them into submission with his voice (303). Cú Chulainn, as Master of the Beasts, is strikingly similar to the constellation of Orion with the three heads as the belt of Orion and the animals as the other constellations in the winter sky. Presumably, he is thought of as the hero of Celtic mythology who assumes the power of the Great Goddess of the Winter Sky.

Although he assumes the power of the goddess at his first triumph over his enemies, Cú Chulainn is bestowed with the power of the goddess at his conception. Much like the heroes of Greek mythology, Cú Chulainn is given the power of the goddess through his ancestry. His father, Lugh, is a solar deity, who falls in love with Deichtine, daughter of Conchobar. Conchobar's lands have been laid to waste by a flock of 180 birds chained by the neck in pairs that lead Conchobar and his company in chase to the Brú na Bóinne at Newgrange. After several adventures with the birds, one of which is Deichtine, Lugh helps Cochobar restore his land and begets Cú Chulainn on Dechtine. This sets a pattern for Cú Chulainn and his associations with the birds later in his life.

In many other Irish-Celtic myths, the hero-god assumes the powers of the goddess. He is often accompanied by a twin and searches for his swan, now in the form of a supernatural being, to bring him love, hope and light. In "The Dream of Angus," Angus Óg falls in love with Cáer Ibormeith, "Yew Berry," and follows her to Loch Bel Dracon with his friend Bodb. At the Loch, they see 150 women, Cáer among them, paired with silver chains. Angus learns that Cáer and the women are swans one year and women the next. Like Zeus, Angus takes on bird form to overcome Cáer as they fly around the lake three times and off to his palace, Brú na Bóinne at Newgrange. The watery element, the pursuit of otherworldly swans, and Newgrange all place this as a solstice myth with the pursuit of love as a dominant theme in Celtic mythology.

The pursuit of love is aptly demonstrated in the "Wooing of Étain" where Midir, ruler of the mound at Bri Leith, falls in love with Étain because of their marriage in an earlier existence. When her husband, Eochaid, refuses to let her go with Midir, Midir places his arms around Étain and they rise up through the skylight in Eochaid's house. When the people rush outside to see Midir and Étain, they see two swans flying around the house three times and then disappearing into Midir's mound. Like Cáer and Angus, the couple shape-change, circle the area and then disappear into the mound of the Otherworld.

In the Welsh legends of *The Mabinogion*, the same love-pursuit theme reoccurs. In the Second Branch of *The Mabinogion*, "Branwen Daughter of Llyr," Branwen is pursued by her brother, Bran, and his company when she is ill-wed to the Irish king, Lord Mallolwch. The names of Bran and Branwen are both associated with white and the winter placing the myth at the appropriate time. Branwen is translated as "white crow" or swan. She keeps a bird as a confident and sends it to Bran in order that he may come rescue her over the seas, a symbol for the watery peril in the myths of the solstice. Finally, the twins in the story, who are cousins of Bran, Nissyen and Evnissyen, also link the myth to the solstice. Bran, of course, rescues Branwen, and the story ends with a reference to Bran and the singing of birds (*Mabinogion* 67-82).

The Irish version of the children of Manawydan fab Llyr, or Manannán Mac Lir, is closer to the original myth of the astral goddess and the twins and has less of love pursuit theme than its Welsh counterpart. In the "Fate of the Children of Lir," Bodb is the king of the

Tuatha Dé Danaan, just as Bran is king in *The Mabinogion*. Lir's children are Aed, fire, Fionguala, white swan, and Fiachra and Conn, the twins. The myth stresses how Lir, the god of the sea, rises at dawn to play with his children and retires at night to his home in the Emain Ablach, the Land of Promise (O'Brien 162). In the tale, the original symbols of the bringer of light, the divine bird-goddess and her twin brothers are evident as is the association with the dawn and the sea. When the children are turned into swans, the aspect of the goddess emerges as a vital part of the myth.

The final appearance of the Great Goddess of the Winter Sky occurs in month of Dumannios on The Sequani Calendar. Here, her third aspect that of birthing, midwifery, and nursing the young is represented by the path of the moon, the constellations and the solar journey of the new year. In Dumannios, the path of the full moon passes in the center of the constellation of Cancer close to the open star cluster of Praesepe, and the path of the sun is marked at the cross-quarter day of Imboc which is a midway point between the Winter Solstice and the Vernal Equinox. On The Sequani Calendar, the full moon is designated by SAMONI in the second, third and fifth years because its ecliptic passes through the constellation of Cancer close to Praesepe on that night. Imbolc is marked on the full moon for the people's celebration or Oenach and on the new moon for the Holy Nights.

Both of these celestial occurrences are celebrated in ancient and modern cultures. The day of Imbolc is celebrated in the Celtic countries on the cross-quarter day, and it is celebrated on the same day in Crete. The second of February is designated as a holiday to St. Brigit in the Celtic countries, and it is celebrated in the cave of Acrotiri near ancient Kydonia in honor of Panagia Arkoudiotissa, "Virgin Mary of the She Bear" (Gimbutas 200). The holidays, converted holidays for Christian saints, were originally dedicated to the goddesses of birth, midwifery, and the young. Likewise, the full moon passing in Cancer near Praesepe evolves from an ancient tradition.

The silvery star cluster of Praesepe is known to many ancient cultures as The Gate of Men. It is said that through this hole descends the souls of human beings (Chartrand 454). Praesepe is the gateway to earth where souls pass from the stars down to earth to be born as humans. In a sense, it is the birth canal of the sky mother's body. Souls pass through the gates of the body of the Great Goddess of the Winter Sky on their journey to earth. Like the star cluster of the Pleiades, Praesepe acts as an opening to other universes and a passageway to celestial time. In Egyptian mythology, it is the golden scarab that holds eggs ready to be resurrected through birth, and in Tibetan mythology, it is a frog with eggs in its belly (Staal 145-47).

The constellation, even as Cancer the Crab, is an image of a being with arms or appendages stretched wide opening at its center to seemingly birth a group of stars, its star cluster, Praesepe. The image is amazingly close to the image of the Great Goddess of the Neolithic Era with upraised arms and parted legs often depicted as a frog, toad or turtle. According to Gimbutas, the votive images of this aspect of the goddess are "intended as a protection against barrenness and to ensure safe pregnancy" (177). A large group of these votives that date from the sixth century BC were found in a sanctuary of Artemis Orthia at Sparta. However, Gimbutas believes the actual worship of this aspect of the goddess can be traced back to the Neolithic Era (176-77).

The associations of the birthing aspect of the Great Goddess of the Winter Sky evolve into two images, the former image mentioned above as an animal figure with the upraised arms and outstretched legs containing the eggs about to transcend to their new universe, and the nursing image of a mother with her newborn young. Both images are clearly represented in the iconography of the Neolithic peoples. The first image is represented in the frog, toad, and turtle votives mentioned above, and the second image is clearly represented in the iconography of the mother and child figurines or Madonna epiphanies of the goddess. Here, the bear is the animal represented as the nurse and protector of the newborn. The "Lady of Vinča" and other Vinča figurines wear bear masks and nurse their young: "Hundreds of such schematized bear-nurse figurines imply the role of the goddess as a protector of weaklings or of a divine child" (Gimbutas 195).

The image of the birthing goddess, the nurse and protector are easily identifiable in Minoan and Greek mythology. In Minoan mythology, she is Artemis Eileithyia, the Artemis who presides over childbirth and is worshipped in caves of Amnissos and Acrotiri in Crete as the "She-Bear." In Hellenic Greece, her epithets are Mogostokos, goddess of the pains of birth, and Phytia, she who stimulates growth. The Tithenidia, the festival of nurses and nurslings in Sparta, honored her name where offerings to her included figurines seated in the birthing posture, and Athenian girls danced to her in the town of Brauronia in Attica wearing masks of bears (Gimbutas 199).

In Greek mythology, Eileithyia is present at the birth of the Eros, the first of the gods to be born from the world-egg, and she is present at the birth of Apollo from their mother, Leto. Artemis is born as the first twin of Leto and then assists in the birth of her brother Apollo. Here,

Eileithyia represents the birthing aspect of Artemis as Artemis Eileithyia. Eileithyia is then present at the birth of Hercules where she and Hera attempt to stall the inevitable event but do not succeed (Graves, Vol. I, 58; Vol. II, 86). Finally, we see her in the myth of Sosipolis where she bears him as her son and he is honored in her temple as the sacred snake (Bell 175). In these myths, Eileithyia is associated with the world-egg, the birth of twins, the sacred snake of the temple, and the hero of the solar deities, all events that clearly connect her to the Winter Solstice as the birthing aspect of the Great Goddess.

The bear and nursing aspect of the Great Goddess is depicted in the winter sky in the mythology of the Hellenic Greeks by Thales of Miletus, Hesiod, and Apollodorus. Here, the image of the bear is used to demonstrate the Iron Age re-naming of constellations to establish Zeus and not the Great Goddess as ruler of the stars. The constellations of Ursa Major and Ursa Minor, now seen as bears by the Greeks, are usually referred to as The Wains by several cultures, both ancient and modern. The British, the Northwestern Europeans, the Babylonians, the Chinese, and even the Romans envisioned the constellations to be a wain, plough, or great wagon drawn by horses or oxen (Staal 121-28). To these cultures, the constellations follow the agricultural cycle as they circumnavigate the north pole marking the position of the plough in relation to the seasons. For example, The Sequani Calendar marks the upright position of The Plough in Simivisonnios at the end of plowing season where the plough is no longer in use.

However, the Greeks using the image of the bear created the myth of Callisto and her son, Arcas, to re-name the constellations establishing Zeus as lord of the skies. In the myth, Zeus seduces Callisto, a companion of Artemis, and she has a son named Arcas. Zeus' wife, Hera, is jealous of Callisto and Arcas and in an effort to save their lives, Zeus changes them both into the bear constellations of Ursa Major and Ursa Minor, respectively. Arcas becomes the ancestor of the Arcadians, immortalized like his mother, and the patriarchy of Arcadian kings is established in myth. The myth emphasizes the bear aspect of the Great Goddess; however, its main function is to demonstrate the changes the patriarchy of Greece imposes on the calendar through politics and myth.

In Celtic mythology, the aspect of the Great Goddess of the Winter Sky as birthing goddess, nurse and protector is faithfully retained in the goddess Brigit. Unlike Greek mythology where this aspect of the goddess is split into minor goddesses, Celtic mythology uses one goddess for the celebration in the winter sky of Dumannios. In Ireland she is Brigit, in Britain she is Brigantia, in Wales she is Brighid, and in Gaul she is Brigantu. The celebration of Brigit is marked as Imbolc on The Sequani Calendar as both an Oenach, or people's holiday on the full moon, and a Holy Night on the new moon. It also incorporates the mid-winter celebration of the cross-quarter days where there is usually a temporary warm spell in middle of winter.

Brigit embodies the concept of midwifery and birthing, as she is protector of the newly born ewes and protector of the newly born ideas of the bards. She is a source of inspiration, represented by a flame or a brightness that burns through the bitter-cold nights of winter. When the first stirrings of life begin in the winter months, she is there to strengthen and protect them. In Celtic mythology, she becomes protector of the tribe as well as the newborn. Her role in Northern Britain as Brigantia is one of "High One" or "Queen" of the Brigantes. Here, she is worshipped by a "strong hegemony of tribes," whose ultimate unity is due to this powerful goddess: "Her name alone suggests her elevated position in tribal tradition" (Ross 453).

The root word of Brigit is the word "briga," an Old Celtic word referring to both light and height, concepts that link the name of Brigit to the goddess as bringer of light in the winter. Her name also reflects the concept of Brigit as more than a local deity. In Celtic mythology, she is a higher level deity who stands over the lesser gods as one of the mothers of the gods (Ross 454). This connects her to the tribes as protector of the land, the cattle herds and the flocks. She is "the goddess of a pastoral people, to whom flocks and herds were of prime importance" (Ross 455). She is described as having been raised on the milk of a white, red-eared cow, the color of the Otherworld beasts. The association with cattle and the yield of the land reinforces her political role as leader of the tribes making her the "High One" in all aspects of their lives.

Brigit's earthly representatives, the Queens of the Brigantes, were bestowed with her power and in a sense, women such as Cartimandua and Boudicca, were a temporal reflection of the goddess. They were her physical embodiment on earth and the Celtic ruler-queens (Ross 454). Like Brigit who keens for her son, Ruadan at the second battle of Magh Tuireadh, Boudicca and the Celtic warrior women experience their enemy firsthand in battle. In this way, both the queens and the goddesses such as Brigit are very different from their Greek counterparts who maintain a submissive position in Iron Age mythology. Brigit's virgin aspect is more than figure of speech because she is an independent and strong figure without a husband or king to rule her. She is often depicted wearing a crown of mural and standing on the globe as a symbol of victory (Ross 455).

Brigit is also protector of the sacred waters of the tribe. As territorial goddess, she is protector of the wells, the springs, and the rivers. Her depiction as nymph-goddess enhances her sovereignty over the land and it presents her as a healer and guardian of the thermal waters; Brigit's "girdle is capable of curing all disease and illness." In Irthington, she is evoked as "deae Nymphae Brigantiae" and in Middlesex and Anglesey, the rivers Braint and Brent are named after her (Ross 455). Her powers are reflected in the rushing rivers and in the hills of the north, epitomizing the strength of the goddess (Ross 461). She is both healer and protector.

Like Artemis, she has a group of virgin priestesses or companions who guard her well and keep her eternal fire lit. Her eternal flame, a symbol of the newly lit fire of the solstice in the darkest winter nights, is combined with her associations of water and healing to bring us to her as a source of winter strength. The image of fire and water as protection, inspiration and the nurturing of new life crown Brigit as the guardian of the new light of the Winter Solstice. The interlude of several warm days in the cold of winter reminds us that life will return to the earth in spring with the help of the Great Goddess of the Winter Sky. It is then that the snake is said to come from the granary, the temple, the earth-mother's body, or the mound to give us sign of his resurrection in the spring.

The Great Goddess of the Winter Sky embodied in her various forms as Brigit, Brigantia, Brighid, Brigantu, Callisto, Eileithyia, Artemis Eileithyia, Fionguala, Étain, Cáer, Dechtine, Helen, Leda, Eurynome, Hecate, Grian, and Artemis are all daughters of the Great Goddess of Life, Death and Regeneration of the Neolithic Era. Her emblems are the constellations, the moon, and the solar events of the Winter Sky. The cycle of the seasons return us each year to the celestial celebration of the Great Goddess as each year we are able to live the birth of creation and the birth of the new light of the year in an everlasting pattern that spirals toward eternity. We embark on a ritual of "eternal return" through the cyclical regeneration of time to transcend time and touch eternity. The stars, moon, and sun are our contact with that eternity and the Great Goddess of the Winter Sky is our guide.

CHAPTER FOUR

The Celestial Spring

Throughout the mythology of ancient cultures the spring months of the Vernal Equinox are a time to reunite with the eternal forces that drive the green fuses of life to the surface. The return of the flowers, the shoots, the sprouts and the young vegetation is represented in the purification of the world by the spring rains which make it possible for the new vegetation to grow to maturity and feed the earth and its peoples. The bitter dichotomy of the new growth that buds in the spring in all the beauty and splendor of new life is that this new life must often be cut, trimmed and sacrificed in order for it to mature and produce fruit. To keep in harmony with this miraculous process, the ancients created rituals in the spring months that are measured in the night sky.

The rituals and celebrations on their calendars follow the welcoming back of the new green life in its infant form, the purification and nursing of the sprouts with water, and the thinning and pruning of the new and old growth for future fruition. We can see this in their creation of deities in the spring constellations. On The Sequani Calendar and in the Lesser Mysteries of Eleusis on the Greek Calendar, the new plants are welcomed by the goddesses who call their young son of vegetation back from his wintry slumber using the tools of agriculture to properly initiate his appearance. This event, known to the Celts as the lunar cycle of Rivros, is marked by the rising of Regulus in the constellation of the Sickle of Leo and by the upright position of Ursa Major, The Plough. The fields are made ready for the new growth and the son is called by his earth mothers to appear.

The purification of the young plant growth, the shoots and sprouts of the earth that rise early in the spring, is dependent on the spring rains that nurse the new vegetation. The ancients saw this as an event of extreme importance because of the fragility of the new growth and deemed that only the goddesses of the calendar year itself be the nurses of the new shoots helping the earth in this delicate process as foster-mothers to the vegetation son. On The Sequani Calendar, the full moon of Rivros passes the constellation of Coma Berenices from which thousands of galaxies and nebulae are viewed from earth. These clusters of eternal time may be viewed as the power of the goddesses. Like the women's festivals in the fall and winter where the full moon passes the star clusters of the Hyades, the Pleiades, and Praesepe, the star clusters of Coma Berenices emphasize the eternal power of the goddess to support and nurse life.

In order for the young plants of the spring to come to fruition, they must be pruned and sacrificed in their prime. To the ancients, this process included the virgin goddess and the young son of vegetation. The virgin nature goddess sacrificed her virginity to ensure fertility and the young son represented by the vegetation is cut or sacrificed for future fruition. This event is marked on The Sequani Calendar as Petiux of the new moon of Rivros and it is marked on the Greek Calendar as the Sacred Marriage of the Spring. In the night sky, the constellations of Virgo, the Virgin, and Boötes, the Herdsman, mark the union of the two fertility deities in the new moon celebration. Virgo, known as Persephone in Greek mythology, spans across the vernal night sky with Boötes at her side.

The sacrifice of the vegetation god is an event that follows the spring marriage and is marked on the calendar of the ancients in several cultures on the Vernal Equinox. On The Sequani Calendar, the lunar cycle of Anagantios marks the Vernal Equinox with the constellations in the west representing the dying god. This aspect of the vegetation god, often depicted in the mythology of the ancients as a bull, is represented as the dying god in the setting of the constellation of Taurus on the western horizon in the second half of the lunar cycle. The sacrificed vegetation god as Taurus sets slowly in the west each night as the raging bull often met by Mars descends below the horizon and is devoured by the setting sun. In the mythology of the ancients, the sacrifice of the vegetation god at the equinox and the death of the Bull on the western horizon are experiences of the same phenomena: the sacrifice of the dying god.

Fortunately, the rites of spring do not end with the death of the vegetation and the demise of the Bull. In the third lunar cycle of the spring, Vega, the first star of the Summer Triangle, appears on the eastern horizon in the month of Ogronios and the full plant life of the summer months begin their maturation process. This star signals the beginning of a new life and resurrection, as it is the singer or musician heralding the arrival of the summer months. The solar event or cross-quarter celebration of Belteine also signals the beginning of the summer with its theme of resurrection through the ascension of the world tree or the plants of the earth.

The celestial journey through the spring has its roots in the culture of the Neolithic peoples. Temples of the Vinča culture settlement in southeastern Europe dating to around 5000 BC reveal an east-west structure seemingly dedicated to a dual-headed goddess or mother-daughter goddesses, most likely a spring-fall pair. In the main hall or eastern upper temple, a life-sized statue of the dual-headed goddess with a bucranium at her feet suggests that the temples, as did most of the Çatal Hüyük temples, were used for rituals of regeneration. On Cascioarele, a Danubian island near Bucharest, a temple dedicated to regeneration from approximately the fifth millennium BC, housed two pillars originally molded around tree trunks which displayed angular spiral designs. A skeleton was found between the pillars and a painted snake hovered above the entrance; the temple walls were painted with red eggs, concentric circles, and snake spiral designs which clearly represent regeneration (Gimbutas *Living Goddesses* 77).

Maltese temples from the same millennium occur in pairs with one temple larger than the other. Gimbutas suggests that these temples were used for the purpose of celebrating the yearly cycles of death and regeneration and that their statues of the dual-goddesses exemplify a mother-daughter twosome similar to Demeter and Persephone in Greek mythology. The smaller temples represent the daughter and face east. In the Ggantija temple, a large ritual basin and a pillar with a low relief of snake carved on it was found. On the wall, a relief of plant spirals and spirals that symbolize cyclical regeneration were also found (Gimbutas *Living Goddesses* 97).

The temples of the Neolithic peoples were most likely used to celebrate the coming of autumn and the coming of spring at the equinoxes. The dual-goddesses, who were later called Demeter and Persephone by the Greeks, were probably worshipped in the temples, their Persephone or daughter figure being the goddess of the east temples which contained the symbols of regeneration and reliefs of the plant life of the spring. The bull, as consort and sacrifice of the vegetation son, is also depicted in the temples beside the altars to the goddess, and the snake relief, a symbol of the snake that was housed in the granary in the winter, emerges in spring and is represented in the Neolithic temples as a powerful force of regeneration. Most interesting, are the angular spirals in the temples. This design, as well as the position and general architecture of the temples, links it to a similar design, which might possibly be the symbol of the equinoxes, on the temples of the Bronze-Age Minoans.

The angular spiral, a symbol of the Minoan labyrinth, is also painted on the Knossos temples in Crete. Minoan mythology uses the labyrinth design to represent the bull god and the rites performed to the god and his mistress, the virgin nature goddess of the spring. In the east section of these east-west temple complexes, the open and airy rooms were used to celebrate the regeneration of spring vegetation. They contain murals with spirals, dolphins, rosettes and an immense statue of the goddess that parallels the goddess in the Malta temples in her life-giving aspect (Gimbutas *Living Goddesses* 136). The labyrinth, the symbols of the goddess, and the presence of the bull aspect of the vegetation god presuppose that the deities of the Vernal Equinox endured beyond their worship by the Neolithic peoples.

In Northern Europe, Knowth, a Neolithic mound in the Brú na Bóinne, also marks the passage of the Vernal Equinox using the angular spiral design on two balains near its entrance. A line, like the line drawn on the center of the balain that divides the solstice spirals at the entrance of Newgrange, splits the angular spirals in half, designating the passage of the setting sun by the shadows cast on to them by two standing stones. The sun's rays penetrate the mound illuminating a cruciform chamber which displays a series of plant motifs characteristic of spring vegetation (Brennan 90). The cruciform chamber, an ancient symbol of sacrifice and resurrection is used here with the astronomical symbols of the Vernal Equinox for ritual use, quite possibly to celebrate the sacrifice of the vegetation god of the spring.

In the Loughcrew mountains in County Meath, Ireland, a massive structure originally including fifty to one hundred mounds extends in an east-west direction on the summit of the range. Dated as far back as the fourth millennium BC, these mounds were constructed as a equinox structure. Inside of Cairn T of one of the remaining structures is a cruciform chamber that is illuminated at the Vernal Equinox (Brennan 68). Like Knowth, the Minoan temples, and the temples in southeastern Europe, the remains of this structure and the remains of many temples from the Neolithic culture demonstrate their use primarily as temples of the goddesses and god of the spring vegetation. Moreover, they mark the celebrations of the Neolithic peoples in a language of astronomy. A calendar stone at Knowth, using the same lunar cycles that The Sequani Calendar uses, and the markings of the solar events on the Neolithic monuments indicate that this culture used the natural cycles of the moon, the stars, and the sun to celebrate their deities.

The celestial bodies are accurate time-keepers of the celebrations of the deities of the Neolithic peoples. The deities of the spring embody the natural events of this time of the year in several celebrations. The first event on The Sequani Calendar and on the Greek Calendar is the calling of the earth mothers to their vegetation son to rise out of the earth where he has been slumbering in the winter months. The Sequani Calendar marks this date as early as the new moon of Dumannios, the second lunar cycle of the year. On The Sequani Calendar, it is the first of five spring celebrations called Petiux. On the Attic calendar of the Athenians, they mark the "calling of Dionysus Limnaios" from the sacred waters of his ancient temples in early spring.

The image of the mother of the vegetation son is found in the mythology of the Neolithic peoples. According to Gimbutas, the Year-God cycle starts with the birth of the divine child. Masked figurines holding and feeding babies or figurines with sacks on the back or with a hump "occur frequently enough to attest a ritual re-enacting of the myth of the birth and nurturing of the Infant" (*Old Europe* 234). In Minoan mythology, the birth of the infant or awakening of the divine child is represented as the newborn vegetation as the Cretan Zeus from the cave of Dikte. He comes "wet and gleaming" from Rhea, his earth mother (*Old Europe* 235).

In Greek mythology, Semele is the name of the earth mother of the vegetation son, Dionysus. Her festival, the Lenaia, most likely took place from the third quarter moon to the new moon of the second lunar cycle of the year as the Greeks celebrated their Holy Nights at this time. To the Celts, this is the beginning of Petiux. According to Walter F. Otto, Semele was represented to the Greeks as a mortal woman who achieved immortality through the birth of her son even though she was worshipped as a goddess for thousands of years by the Phrygians (70). When Zeus reveals himself to her in his full power, she is consumed by his lightning and fire. The destruction and demotion of the goddess as well as her cult is typical of classical Greek influences; however, her associations with the life giving attributes of a cult of mother goddesses named as Rhea, Demeter, and even Persephone retain some elements of her stature as Divine Mother.

The most interesting aspect of the care and nursing of the vegetation son is the fact that his mother immediately gives him to other goddesses to be nursed. With the celestial events of the upright position of the Plough and the rising of Regulus in The Sickle of Leo on the eastern horizon, the next phase of the celebrations of spring indicate the care and nurturing of the new vegetation as the central focus for the people. The advent of the full moon in the beginning of the third lunar cycle of the year begins as Ocio Rivros and ends as Devo Rivros. This lunar cycle initiates the growth of the new vegetation and is carefully marked in the nine holy nights around the full moon. The ecliptic of full moon at this time passes the constellation of Coma Berenices. On the Greek calendar, this is the time of the Lesser Mysteries of Eleusis.

The nurses and foster-mothers of the vegetation god are an important aspect of the Neolithic pantheon. They are the spirits that immediately care for the young god and their most important task is to nurture the young god. According to Gimbutas, masked goddesses in the shape of a bear, snake or bird nurse the baby as cub, snake or baby bird (*Old Europe* 234). In later Minoan and Greek mythologies these are Artemis as the bear or great goddess of the winter, Demeter as the great goddess of the fall with the snake that protects her granary, and Athena as the bird and snake goddess of the summer. Each of the seasons is therefore called upon to aid in the nurturing of the spring vegetation. According to Gimbutas, a sanctuary of Dionysus on the island of Keos dating back to the fifteenth century BC, revealed "figurines portraying women in a dancing posture, dressed in festive attire, with exposed breasts, snake collars, and belts (*Living Goddesses* 162). Indeed, these might be some of the first representations of the goddesses that nurture the vegetation to grow with their ecstatic dance.

In a hymn of invocation to the divine child engraved on a stone in Palaikastro, Crete, the foster-mothers are referred to as "the shielded Nurturers" (*Old Europe* 235). In Mycenaean mythology, they are called the Couretes who perform their dances around the vegetation son, Dictaean Zeus. Their dance is both a fertility dance and a dance of protection; perhaps they are the ones in the hymn who "leap" for fertile fields and full jars. The dual role of foster-mother or nurse and dancer are two distinct ways of depicting the role of the goddesses that surround the vegetation son. In one way, they must passively nurture and protect the infant vegetation and in another, they must stimulate his growth with an active fertility dance.

A third aspect of the goddesses that nurture the vegetation god is probably their most important and that is the role of sustainer of life through water. The bringing of water to the fields is the sacrifice and difficult task of the women in the agricultural cycles of the spring. Without water, the vegetation will not grow and without the work of bringing the water to the fields, the plants will not grow. The goddesses of the spring are associated with water when they first discover the divine child emerge from the watery depths at the sacred springs of Lerna or at the Athenian temple near the ancient springs of the Acropolis. Sacred waters from these symbolic springs are the holy waters used in the Dionysian rites that mix the wine with water. The goddesses or their earthly representatives, the priestesses, are in charge of the holy waters for the spring vegetation.

In the Lesser Mysteries of Eleusis, water as a purifying element is used in the ceremonies for the initiates (Mylonas 242). In the festival of Agrionia of the Lesser Mysteries, performed in the full moon of the third lunar cycle, the initiate must be cleansed in the river with her/his sacrifice of a sow to Persephone, the virgin goddess of the spring, before she or he receives the initiation rites. She or he is then brought to the temple and a liknon or winnowing fan is placed above her/his head as a symbol of restoring the seed of the new year. She or he is then received by the goddess Demeter in her preparations for her Greater Mysteries in the fall. The purpose of this ancient ritual is to purify and cleanse the initiate in much the same manner that the young vegetation is prepared and watered in the spring by the spring goddesses.

The representative of the goddesses who prepare, water, care and nurse the young vegetation is the constellation of Coma Berenices in the night sky. Coma Berenices is a constellation that is associated with the goddess. It was named in third century BC after Ptolemy II's wife, Queen Berenice who sacrificed her hair to the goddess if her husband would return safely from battle or the world of the dead. Patterned after the myth of the goddess Isis, who sacrifices her hair in order that her husband, Osiris, return from the dead, Berenice is aptly deified in ancient Egypt. The theme of sacrifice to bring back the male god from the realms of the dead finds its agricultural metaphor in bringing back the vegetation god from winter.

The shape of the constellation and the position of the constellation to Virgo, the virgin goddess of the spring, make its relevance apparent. It is shaped much like the winnowing fan or liknon used in the Lesser

Mysteries and it is positioned near the head of Virgo; the liknon was carried on the head of the priestesses in the ceremonies (Nilsson 31). As a symbol of the purity and sacrifice of the goddesses of spring, Coma Berenices is aptly marked on the ancient's calendar when the full moon passes it in the third lunar cycle of the spring. In the opening of the fan of Berenices, thousands of galaxies and star nebulae give the stargazer a view of "the regions beyond our own galaxy" (Chartrand 490). This view of eternity symbolizes the immortality of the goddesses that guide us through time.

In the Dionysian Mysteries, the goddesses or priestesses place the vegetation son in the liknon in the form of a veiled phallus with fruit. The return of the god as phallus emphasizes the fertility aspect of the spring rites; however, according to Vergil's *Aeneid*, the spiritual metaphor is not lost. The "new" conception of the world, popularized in the Hellenistic age, transferred the Underworld into aerial space and ascending souls are said to be purified by the liknon as they were tossed in the air as chaff (Nilsson 370). The ascendance of these souls to the galaxies as the chaff or debris in Coma Berenices illuminates both the spiritual and agricultural meaning of the constellation. The goddesses as the liknon, raise, nurture and present the fertile powers of spring and take care of the souls who have been purified in their rites by placing them in eternity in the night sky.

The emphasis of the vegetation son in the Dionysian Mysteries as phallus changes the role of the goddesses as the nurturing forces of the eternal return of spring vegetation to frenzied women called Maenads. Although their names change in reference to the specific cults of Dionysus, such as the Dionysiades, the Bassarai or the Klonones, they are the fourteen women, the Gerairai, who mix the wine with the holy water and perform the ecstatic dances (Otto 175). They are foster-mothers and nurses who live in forests like wild beasts and suckle the young of wild animals as if they were their own. Unfortunately, they have become the frenzied ones who tear their young to pieces in a fit of madness. Perhaps later, in Indo-European cultures, their ecstasy became madness and was often accompanied by punishment. The use of the phallus and the initiation of the wine culture change the mysteries significantly.

The change of the goddesses to Maenads is indicative of the many myths that accompany the arrival of Dionysus in Greek mythology and the change to a patriarchal form of government. The resistance to the new cult of Dionysus is evidenced in the violence of the myth of Lykourgos where he grabs the Minoan axe from the women who perform the sacred rites of spring and kills them. The women who hide the divine child in the festival at Boeotia are also chased by the priest of Dionysus and the one he catches is killed. Likewise, in Hera's cult at Argos, the College of Sixteen Women fiercely resist the change to the new Dionysus (Kerenyi 176-185). All these myths have the same foundation of ritual persecution of the women celebrants and subsequent madness as punishment stimulated by the patriarchal takeover of the ancient rites of spring

In the second stage of the third lunar cycle of the year which takes place in the new moon of Rivros on The Sequani Calendar and in the new moon of Anthesteria on the Greek Calendar, the Sacred Marriage of the virgin nature goddess to the vegetation son takes place. On both calendars this is marked by the three nights sacred to the new moon. Where The Sequani Calendar marks both Ocio Rivros and Devo Rivros as the first stage of Petiux, the Greek calendar marks the Agrionia as the first stage of the Dionysian rites. The second stage on both calendars occurs in the new moon and again on The Sequani Calendar is marked as a festival of Petiux in all five years of the calendar. In the Dionysian rites, it is the Sacred Marriage nights in month of Anthesteria.

In the night sky, the constellation of Virgo represents the nature goddess of the spring. The constellation appears on the eastern horizon in the new moon of the third lunar cycle and spreads across the night sky as one of the largest constellations lying across the moon's ecliptic. In many ancient cultures, Virgo is the goddess of the spring who brings fertility to the land when she marries the year vegetation god. In Babylonian myth, she is Ishtar, Queen of the Stars, in Egyptian myth she is Isis, and in Greek myth, she is Persephone. In her constellation, she is seen holding a spear or spike of wheat or grain as a symbol of her association with agriculture; this spike is marked by the primary star of Spica. Her appearance in early spring indicates her yearly renewal and her release from her wintry realms as goddess of the Underworld. In this pattern of eternal renewal, her role as maiden and seed is to unite with the vegetation god in order that the vegetation may begin to grow and flourish.

In the Neolithic Era, the goddess appears as daughter to the Pregnant Vegetation Goddess where each year she is stored in the temples as the seed and restored to the land in the spring; the daughter dies annually and the grain mother mourns her death. In this way the cycle of regeneration and germination would be enacted on a yearly basis. A sacrificed pig was used as the symbol of the daughter in both the fall and spring rites and pithoi jars with samples of grain were used in the temples to store the sacred grains of the daughter. When the daughter returned in the spring, her sacred marriage was celebrated. The sacred marriage of a divine vegetation pair is depicted in figurines as early as 10,000 BC in the Near East and in 5000 BC in southeastern Europe. In the written texts of the Near East, descriptions of the hieros gamos or sacred marriage also refer to a coupling of the goddess, or priestess, and her consort, the god or priest, who performed sexual intercourse for the well-being and fertility of the land (Gimbutas *Living Goddesses* 18-19).

According to Gimbutas: "The iconography of the sacred marriage is recognized in frescoes, gems, and ivory reliefs found in Indo-European Mycenaean and later Greek art. The repeated theme is that of a bridegroom who comes in his ship from overseas to fulfill the sacred wedding in the sanctuary of the goddess" (*Living Goddesses* 118). In the Ship Fresco of the West House of Akrotiri on Thera, the goddess welcomes the bridegroom to her own temple for the sacred marriage amidst the blooming flowers of spring thus preserving what Gimbutas calls "the matrilocal tradition." These earlier traditions represent the goddess of spring as a complementary partner in the sacred marriage in a role similar if not identical to the male role.

Greek mythology tells the story of a Minoan princess who is immortalized as the goddess Ariadne because of her sacred marriage to the vegetation god. Ariadne marries Dionysus on Naxos where an elaborate dance ceremony called the Geranos is performed in her honor. Ariadne, known as "Mistress of the Labyrinth," watches as the dance performed for her follows the pattern of the labyrinth. The dancers perform the dance in much the same manner that actual cranes perform their mating dance in the spring. Ariadne is their supporter, their priestess, and their goddess. Like Persephone, Ariadne bears the epithet of "utterly pure" perhaps meaning that she could appear "utterly clear in the heavens" (Kerenyi 104).

On coins from Knossos, her head is framed with the square meander pattern of the labyrinth or the angular spiral found on the Neolithic temples of the Vernal Equinox. The angular spiral is often accompanied by astronomical signs such as the waning and waxing moon and a star inside the labyrinth (Kerenyi 105). In one version of the sacred marriage myth, Ariadne and Dionysus ascend to the peak of a mountain on Naxos and disappear into the heavens (Otto 181). Ariadne's marriage to Dionysus puts them both in the night sky as an immortal symbol to their union. Ariadne is the virgin goddess of both the sacred marriage and the celestial spring who is glorified in the night sky to celebrate her return from death or winter.

The legends of the Greek heroes depict the dance of the labyrinth in the myth of Theseus and Ariadne as one associated with the Dance of the Cranes. Here, Ariadne remains the princess of the island and does not become a goddess. It is Theseus' role to marry her and take her away from Crete to mainland Greece as a symbolic gesture of the Hellenic Greeks conquering of the Minoan empire; however in some myths, at the island of Naxos, Ariadne is abandoned by Theseus and is found by Dionysus who instantly falls in live with her and marries her thus immortalizing her in the night sky. In other versions of the myth, Ariadne hangs herself or dies in childbirth and then ascends to the heavens as the wife of Dionysus who places his crown in the night sky for her as the constellation Corona Borealis (Graves, Vol. I, 336-340).

The death of Ariadne might possibly have its origins in the Neolithic artifacts that portray the deaths of two vegetation deities, a male and a female, found in a Cernavoda grave of the Hamangia culture in Romania dating from the fifth millennium BC (Gimbutas *Living Goddesses* 145). As vegetation goddess, she must die or descend to the Underworld and return in the spring annually accompanied by her consort. According to the Homeric Hymn to Demeter, Ariadne came to Greece from Crete: "At Naxos, Ariadne had two festivals, one of them celebrating her marriage and the other mourning her death" (Gimbutas *Living Goddesses* 145). Her death, as comparable to the descent of Persephone to the Underworld, might have been a natural part of the agricultural cycle in ritual and myth in earlier cultures. However, her death in the Greek myths is violent or self-imposed compared to a more passive return to the Underworld for the winter as goddess of vegetation.

Likewise, Persephone's marriage in Classical Greek mythology to Hades, the Lord of the Underworld and a representative of the chthonic aspect of the Vegetation Year God, is a violent abduction symbolic of the domination of the Greeks over the earlier cultures. In Greek mythology, the Vegetation Year God, like the goddesses of the Neolithic pantheon, is represented by separate deities. Where Dionysus represents many of the aspects of the Vegetation Year God, one aspect, his chthonic association, may be represented by Hades in the mythology of the Greeks. In the myth of "The Dethronement of Cronus," the three new rulers and brother-kings, Zeus, Hades, and Poseidon hold a war counsel to determine the division of creation into three realms to be ruled by each of them (Graves, Vol. I 40). As both the culture and the seasons are dominated by these patriarchs, Hades is given "the helmet of darkness" to rule the underworld, an apt representative of the chthonic forces of creation formerly identified with the Vegetation Year God who descends to the underworld for the winter. Persephone's marriage to Hades, unlike Ariadne's sacred marriage to Dionysus, is a death-like term of confinement and a quiet resignation to the determinations of the new patriarchy.

Fortunately, Ariadne's marriage to Dionysus at the end of the Theseus myth retains the celebration and vital renewal of the sacred marriage in the spring of the virgin goddess with the vegetation deity because the sacred marriage is of great importance to the culture; however, it too undergoes some changes. In the rituals, festivals, and celebrations in the Dionysian Mysteries, the sacred marriage is celebrated in the spring. The Dionysian Mysteries first celebrate the Lesser Mysteries in the Agrionia on the full moon where initiates of the Mysteries of Demeter and Persephone are cleansed and purified with the sacred waters of spring. A pig is sacrificed to Persephone and the initiate is cleansed by the foster-mothers and goddesses or priestesses of the calendar year. However, in the new moon celebration, the role of the virgin goddess and her consort in the sacred marriage, although still represented, undergoes some changes.

On the Attic Greek calendar in the new moon of the month of Anthesteria, the Athenians celebrated the sacred marriage of Dionysus to the wife of the official who had the title of the archon basileus. Fourteen Venerable Women called the Gerairai act as the priestesses of the chosen wife. On the eleventh, twelfth, and thirteenth of the month, those days which follow the new moon, the pithoigia or jars of wine are opened in order for the god of wine, Dionysus, to ascend. The Day of Pitchers, the twelfth, is the sacred day to mix the wine with the holy waters, and the thirteenth, or Day of the Pots, the celebration of the swinging of the virgins, symbolic of the hanging of Ariadne, takes place. The wife of the archon basileus is then led in a marriage procession to the Boukoleion, or bull's stall, to marry Dionysus as the Bull (Kerenyi 301-310). Here, the role of the goddess merges religion with politics.

The role of the virgin in Greek mythology was influenced by the ideology of the Indo-Europeans. According to Miriam Robbins Dexter, virginity was seen by the Indo-Europeans as energy which was an unused potential that was stored by the woman while she was a virgin. In a male-centered society, "it is the state which renders a woman a 'storehouse' of untapped energy, energy which she will, in her next stage, impart to man." The virgin, a "repository of energy," was able to transmit her powers to man without diminishing those powers, thus retaining her virginity perpetually (Dexter 161). The wife of the archon basileus, whose virginity was most likely restored symbolically through ritual, became the representative of the goddess, and was therefore able to participate in the celebration of the sacred marriage. She gives herself to the bull god in order that the patriarchy may benefit from the transmitting of her energy.

The role of the vegetation god has also changed significantly. Instead of god of vegetation, he is god of wine and vegetation. His aspect as the vegetation son where the goddesses nurture him is overshadowed by his presentation as phallus in the liknon that the priestesses carry on their heads in the marriage procession. Moreover, his role in the sacred marriage is one of sexual excess and violence. The bull aspect of Dionysus can be traced as far back as the Neolithic Era where it is a symbol of the fertility of the agricultural year god; however, his role of bull-man in the Dionysian rituals is predominantly sexual. His aspect as the dying god, which can also be traced back to its Neolithic roots, becomes violent in the Dionysian rites. In the Dionysian Mysteries, his appearance in the spring has become one of a raging bull about to be sacrificed after his marriage in the Bukoleion.

The associations of the vegetation year god with the bull are evident in the Neolithic culture as symbols of fertility and growth. The role of the bull, often depicted as half-man and half-animal, represents the importance of regeneration in an agricultural community. According to Gimbutas, the half-animal, half-human aspect "reaches a culmination of power through symbiosis: the wisdom and passions of man merged with the physical strength and potency of the bull" (*Old Europe* 224). Gimbutas states that Dionysus' origins belong to the Mother Culture: "Dionysus was a bull-god, god of annual renewal, imbued with all the urgency of nature" (*Old Europe* 227).

The Bronze-Age Minoans presented the bucrania or bull's horns found in many of the Neolithic temples including those of Çatal Hüyük as an important image of regeneration in their culture. In the Cretan festivals, images of the bull as representative of the year-god are prevalent on the seals and artifacts (Kerenyi 116-117). Unfortunately, the Cretan bull in the myths of the Greeks becomes the Minotaur of Crete, a monster that is born from a bull sent by Poseidon, the god of the sea, and Pasiphae, mother of the mortal Ariadne and queen to Minos. Like Zeus who changes himself into a bull to abduct Europa, the mother of King Minos, the bull is depicted as a sexual monster symbolic of a tradition that must be conquered. In the same vein, the Minotaur, half-man and half-bull, still a powerful symbol of fertility and strength, must be conquered by Theseus; Minos' chief captain, Taurus, must also be conquered by Theseus. The Minotaur, a product of the lust of Pasiphae for the bull, becomes the bull god revealed as a monster. His lust for women and his sexuality are revealed as a character trait that must be extinguished by the gods and heroes of Greek mythology.

The bull of the Dionysian Mysteries is another bull of sexuality and violence. It is he who is born with horns and torn apart by the Titans when he manifests himself as Zagreus. As Dionysus Laphystios, he is worshipped by women who wear horns and the women of Elis invoke the god with his bull-foot. Among the Argives, Dionysus is called the Bull-born (Gimbutas *Old Europe* 227). In the sacred marriage, he is the sexual bull-man and in The City Dionysia, a festival concurrent with the sacred marriage, in the procession outside the Bukoleion temple, "the image of Dionysus, possible in bovine form, or an actor wearing horns and a hide, was brought on a boat-like structure on wheels to complete the nuptial rites" (*Old Europe* 228). Here, as in his myths of spreading his religion, he arrives as the conqueror.

When Dionysus completes his marriage with the wife of the archon basileus, he is sacrificed and violently torn apart. This violence, which led to the rise of tragedy as a dramatic genre in the Dionysian Mysteries, is depicted in the night sky as the setting of the constellation of Taurus, the Bull. This stellar and lunar drama is complemented with the solar event of the Vernal Equinox. On The Sequani Calendar these events mark of the beginning of the month of Anagantios and on the Greek calendar, the beginning of the month of Elaphebolion. As Taurus sinks slowly below the horizon in the west, he is devoured by the sun's rays in a violent death. He is pursued by Mars, the red planet of war and death, as he descends into the waters below the horizon.

The Dying God is an archetype that has evolved from the death of bull aspect of the vegetation year god in the Mother Culture of the Neolithic peoples. The death of the spring deity was supposed to bestow upon its participants the power of the god who had been killed (Otto 107). In ancient cultures, a bull or he-goat is sacrificed at this time of the year. In Europe and the British Isles, he was worshipped as Tarvos, Esus, Teutates, and Taranis; in ancient Phrygia he was worshipped as Attis; in Syria, he was worshipped as Adonis; in Egypt, he was worshipped as Osiris; in Phoenicia he was worshipped as Ba'al; and in Greece, he was worshipped as Dionysus and Hyacinth. The

Indo-European culture's emphasis on The Threefold Death, influenced the worship of this deity significantly perhaps contributing to the wide distribution and popularity of the image of The Dying God.

In the Indo-European culture, The Threefold Death represents the trifunctional division where the god is sacrificed by three different means each representing a function of the society. He is sacrificed for the sovereignty, for the warriors, and for the increase of the tribes and their fertility. In each of three deaths, he is connected to a specific functional identity. For example, he is hanged for the sovereignty, he is slain with a weapon for the warriors, and he is buried alive or drowned for the commoner. The Dying God, or King as his substitute, is sacrificed for the good of the tribe. According to Dean A. Miller, "the triple-death theme spreads downward or outward from royal or heroic legend, and is widely current in the folklore of Indo-European speakers" (578).

In the eighth century BC, the Archaic Greeks named the constellation of Boötes as a symbol of The Dying God. In Greek mythology, Boötes or Icarius of Attica is known as the herdsman, the ploughman, and the inventor of wine. Like Dionysus, he dies a tragic death. Dionysus teaches Boötes how to make wine and at a feast of Dionysus, Boötes is killed by his shepherds. When his body is found by his daughter, Erigone, she hangs herself. Boötes is placed in the night sky as a constellation and his daughter is placed beside him as the constellation of Virgo. The myth parallels the life and death of Dionysus and the constellations follow the pattern of the story. The primary star of Boötes, Arcturus, rises in the east as an indication that the sacrifice of the dying god is happening. Arcturus, a very bright yellow-orange star, is an appropriate symbol of the sacrifice of The Dying God; it too marks the beginning of Anagantios and Elaphebolion. Moreover, it complements the tragedy of the death of Taurus on the western horizon.

On The Sequani Calendar, the sacrifice of The Dying God is marked as Ocio MV Rivros, the fifth and final celebration of Petiux. To the Celts, this happens at the same time as Mean Earraigh or the Vernal Equinox. The sun's rays will be longer after the equinox as an indication that the god has died for the people's crops to grow and flourish with the strengthening rays of the sun. The Sequani Calendar and the Neolithic monuments of the Celts support the fact that the Vernal Equinox was celebrated in the life of the ancient Celts. The calendar and monuments carefully mark the events of the celestial spring and the mythology of the culture reflects the same themes that the Minoans and Greeks depict in their cultures. In the myths of the Celts, the dying god is of central importance and the preparations for his arrival, his sacred marriage, and his death are a vital part of Celtic culture.

In Gaul and Northern Britain, the mother of the vegetation god is represented as the Matrona, the Matres, or the Three Mothers. The Matrona are often depicted as three women seated on a stone bench holding cornucopias, fruits, bread, or an infant. In Ireland, Anu or Danu is the mother of the tribes, and in Wales, Modron or Don, is the mother goddess as "the waters of heaven" (Ellis 83). As divine mothers, the Celtic mother goddesses, like the Greek goddesses of the spring, are associated with water as a sacred source of veneration and life-giving power. According to Anne Ross: "Rivers are important in themselves, being associated with deities such as the divine mothers and the sacred bull" (46).

The Matrona in their three-fold aspect embrace the concept of the mother and the nurses as the goddesses that surround the divine child in the mythologies of the Neolithic, Minoan, and Greek cultures. They are the source of his being and the waters that nurse him into existence. They are associated with the sacred waters of well sites, rivers, and lakes which are the entrances to the Otherworld in Celtic mythology. As goddesses of the sacred waters that birth and nurse the divine child, they represent fertility and strength. Often, the river goddesses engage in ritual mating with gods at their sacred waters with a foot on either bank. As fertility figures, they support the life forces of the vegetation god of the spring as their son.

Like the Celtic myths of the swan in the winter sky, the virgin nature goddess of the celebrations of spring is represented by a sacred water bird: the crane. The crane-heron, a form adopted exclusively by females in Celtic mythology, is a bird goddess depicted in a bronze figurine from the temple of Maiden Castle in Britain and on a relief from Notre Dame in Paris on the back of the bull as three women or cranes (Ross 363). According to Ross: "Its appearance as a female form of metamorphosis in the insular traditions suggests that it was originally the attribute or shape assumed by some potent and formidable goddess, consort or servant of a fundamental god-type, Esus, Midir, Manannan" (365). Most likely, the goddess Ross refers to is the virgin nature goddess of spring, consort to the vegetation son or dying god and associate of the water deities of spring.

In the mythology of the Irish Celts, the goddess appears as Aoife, wife of Manannán Mac Lir, god of the sea. She is a foster-mother of

Lir's children and a goddess from whom the sacred crane bag was made. In the legend of the crane-bag, Aoife and Iuchra fall in love with the same man, and jealous Iuchra turns Aoife into a crane while they are swimming in the waters of the sea god, Manannan. Aoife must remain in bird form for two hundred years dwelling in Manannan's watery regions and eventually dying. At Aoife's death, Manannan makes a sacred bag of her skin to hold precious tokens symbolic of his powers that he eventually bestows upon the heroes. Aoife is associated with water as a regenerative power in the form of her husband, Manannan. Her sacred marriage to the god of the sea also gives her the ability to bestow her powers of purity, beauty and longevity to the heroes of Celtic myth.

In a similar Irish tale, Miadhach, daughter of Eachdhonn, is met by the hero, Oisin. She tells him that she was once a woman who has been in crane form for two hundred and ninety years because of a jealous fight with another woman over a young man named Abharthach. Miadhach tells Oisin of all the animals that have been her lovers over time. Although she does not die and achieve immortality like Aoife, Miadhach lives to be one of the oldest animals. Like many of the cranes in Celtic legend, she lives a long life as a solitary figure. In the Welsh *Mabinogion* when Mapon, son of the mother goddess, Modron, is rescued from his stone castle beneath the waves, he is guarded by the oldest animals. The longevity and immortality of the deities associated with the crane as one of the oldest animals reflects their powers of regeneration and renewal. In Celtic mythology, the goddess' role is to protect, guard and sustain that sacred power.

The crane goddess as protector and consort of the Otherworld deities is part of a complex of images which includes the bull, the cranes on his back, and a willow tree branch felled by Esus (Ross 351). This is clearly a complex that represents the sacred marriage of the bird and the bull, an ecological phenomenon present in the bird and bull's symbiotic relationship in their watery marshes. According to Ross, this bull-tree-wading bird complex dates from the Urnfield and Hallstatt cultures onwards. Esus, as the dying god and a god of war, is depicted by Mars in the Gallo-Roman period and the bull is replaced by the horse with cranes on his back. The role of the crane in this complex eventually became one associated with protection and war (Ross 353).

Ross makes an interesting statement concerning the image of the crane goddess in the mythology of the Celts. According to Ross, the later myths of the Celts take the image of the goddess, depicted in the myth of Aoife, as a gentle creature, a goddess truly representative of the eternal regenerative powers of life in the spring and modify it into a mortal woman who is involved in jealous relationships. In the myth of Midir in the Book of Leinster, the crane becomes a frightening creature. Although she is seen as a protector in times war, she is also "symbolic of parsimony and unpleasantness" when in her three-fold form she guards the mound of Midir, Bri Leith (356). Ross states: "Although clearly venerated in the earlier period, the bird seems to have developed a quality of mystery in the popular mind, and to have inspired a certain amount of fear and dislike in the minds of the Celtic peoples" (Ross 369).

The evolution of the crane-goddess to fierce protector and woman to be feared is one much like the evolution of the goddess in the myths of the Minoan and Greek cultures. The Minoan dance devoted to the Mistress of the Labyrinth, known as the Geranos, a word meaning "crane" in Greek, represents the goddess in her most admirable form in the ceremonies of the sacred marriage. The connection of the cranes with the goddess is faithfully represented in the Gerairai of the Dionysian rites who are the fourteen priestesses that serve the wife of the archon basileus in the sacred marriage to the bull. However, in the myths of Theseus and Ariadne, the cranes are fourteen mainland Greek youths who must perform the Dance of the Cranes with the bull and then be sacrificed; Theseus, of course, must end this Minoan ritual of horror against his people. Likewise, the Iron Age mythology of the Celts, associates the crane with war, greed and evil foreboding.

Fortunately, some evidence of the original love dance in the sacred marriage between the goddess of the spring and the vegetation god is found in the celebrations of two rituals of the Celts. One is the Troy-town dances performed in the turf-cut mazes of Britain where a sacred egg is found as a symbol of new life in the spring (Graves.Vol. I, 370). The other is the dance around the fire in the first celebration of Petiux, known as the Escougion in Belgium and France. Here, fires are lit in honor of the beginning of the spring rites. Young men and women run through the orchards and fields with torches of fire to purify the crops. The couples then gather around a fire and when the flames blaze up, they leap over the fire. Dances and songs chanting to "corn in the winnowing-basket, the plough in the earth" are performed around the fire to ensure the earth and its peoples with abundance in the coming months (Frazer 696-701).

The role of the vegetation son in the sacred marriage of the spring in Celtic mythology is similar to the role of the vegetation son in Minoan and Greek mythology. His three aspects as the divine son of the mother goddess, the bull, and the dying god are faithfully represented in the myths of the Celts. As the divine child, he is Mapon ap Modron in the Welsh legends of *The Mabinogion*; Maponus, son of Matrona, goddess of the Marne River in France; Maponus in Northern Britain and Gaul; and Mac ind Óg, son of Madron in Ireland. As the bull, he is Tarvos Trigaranus, "The Bull with the Three Cranes" in Gaul and the Brown Bull and White Bull in the *Táin Bó Cuailnge* of the Irish. His appearance as the dying god is represented in the triskele of Esus, Taranis and Teutates where he dies the Threefold Death found in Indo-European ideology.

In Welsh mythology, Mabon ap Modron is a surviving figure of the son of the mother goddess. In "How Culhwch Won Olwen," a tale from *The Mabinogion*, Mapon is rescued from his captivity at Cáer Loyw, a synonym for the Otherworld, where he has been since he was stolen from his mother three days after his birth. Like Dionysus, he is taken from his mother and appears from the sacred waters of the Otherworld where he has been kept. As a powerful force of regeneration, Mapon must be brought back to earth. When his friend, Culhwch, comes with Arthur and his men to rescue Modron, Modron repays Culhwch by helping Culhwch to fulfill the tasks required for Culhwch's marriage to Olwen, a princess that charms all with her outstanding beauty (*Mabinogion* 163-176).

Although the marriage is not his own, Modron is involved in a marriage with sacred vows and acts as a brother to Culhwch. Modron helps Culhwch kill Twrch Trwyth, sacred pig of the Otherworld, as one of his tasks. His associations with the pig and his return from the Otherworld to aid in a sacred marriage are comparable to the sacrifice of the pig in the Lesser Mysteries of Eleusis and the sacred marriage in the celebrations of spring in Greek mythology. Although *The Mabinogion* is the only surviving text of the "Divine Youth," a spring dedicated to Maponus was found in Gaul and five inscriptions dedicated to him in Northern Britain attest to his widespread cult among the Celts (Ross 464). In Ireland, he is known as Mac ind Óg or Aonghus, the son of Madron.

In 197AD, Strabo writes in his *Geographia* that the Celts are a people that believe men's souls and the universe are indestructible forces (Ross *Druids* 15). Like many foreigners of his time, Strabo was fascinated with the Celtic belief in the immortality of the soul and the metempsychosis of the spirit from one form to another through eternity. In their religion, the Celts believe that the transformation of spiritual energy whether in human or animal form moves through time retaining its memory and purpose. In the Welsh *Mabinogion*, the spirit that guards Mapon in his underworld abode must undergo several transformations in order that the regenerative spirit of Mapon be kept alive in the Otherworld. Likewise, in one of the earliest Irish Celtic myths, *The Book of Conquests*, the spirit of the tribe, Tuan the White Ancient, transforms after his dream-time sleep of death into a stag and then a sea-eagle in order to retain the memory of his ancestors with each successive tribe that conquers the island. In the "Song of Amairgen" in the *Leabhar Gabhala*, the poet undergoes metempsychosis to retain the spirit of the bard.

In the myths surrounding the divine bull, metempsychosis is a dominant theme suggesting an emphasis on regeneration typical to the myths of the spring as a natural function in many cultures where the vegetation cycle is a basis for their ideology. The bull, as a figure of regenerative power can be traced to the Urnfield period with its connection to solar cults. In the iconography of the Celts, the bull is seen as Tarvos Trigaranus, "The Bull with the Three Cranes" and the three-horned bull of Gaul, Northern Britain, and Belgium. As a sacrificed divinity, the bull is also seen in the form of the bucrania on ritual buckets and vessels from Britain and in a ritual slaying scene on the Gundestrup Cauldron (Ross 384-6). Here, it represents the power to sustain life immortal through sacrifice, transformation, and regeneration.

In the Iron Age myths of the Celtic heroes, the bulls of the *Táin Bó Cuailnge* are a powerful image of the bull as regenerative energy. The tale begins with two pig-keepers arguing about which one of them is the most powerful. The pair challenge each other in test after test as birds of prey, undersea creatures, stags, warriors, phantoms, dragons, maggots, and finally, as the dark bull of Cuailnge named Dub and the white bull of Ai Plain named Finnbennach (Kinsella 46-50). The two spirits are the archetype for the warrior spirit in Celtic mythology. In each transformation, they are equal combatants, powerful, ferocious, and strong in their efforts to overcome each other. In the end, the Brown Bull kills the White Bull and then perishes as a result of the effort needed to kill his opponent.

To the Iron Age Celts, the regenerative power of the Brown and White bulls is transformed into the spirit of warriors to reflect the strength of the warriors who face death and transcend the fear of death with the knowledge that they will come back as warriors in another form. The emphasis on the regenerative forces of the warrior in connection with the bull is demonstrated in the Celtic tarbfeis or bullfeast, a ritual of the Druid priests. In the tarbfeis, a bull was slain and cooked; its "broth was then bathed in by the presiding priest and its flesh eaten by the priest and his assistants, and in the sleep which ensued, the king elect would be revealed to the priest" (Ross 387). In this ritual, the belief in the transformation of the bull's energy and power is given to the king-elect as the most powerful of the warriors through the sacrifice of the bull, and the priests are the mediums through which the energy of the bull is successfully transformed.

The bull as warrior, king and bull-lord has its roots in the Indo-European warrior elite represented in the myths of the cattle-raid. Here, the cattle are stolen by a non-Indo-European enemy people and are taken by force by an Indo-European warrior elite. The cattle are then returned by the victors to a First Function priestly class to be sacrificed to the goddesses and gods. According to Miller, the cattle-raid as a warrior activity is "widely known and deployed throughout our Indo-European sources; here it is sacralized and even cosmicized." Seen in the Indic royal consecration as the rajasuya, it is featured in the *Iliad*, *The Odyssey*, and the *Táin Bó Cuailnge* (635).

Sacrifice plays a vital role in the bull cult of the Celts and in the worship of Esus, Taranis, and Teutates, a triskele of gods who represent the vegetation son as The Dying God. This Celtic triune of gods represents the Threefold Death of the Indo-European culture. Esus, is the hanged god for whom the sovereignty were sacrificed; Taranis is the thunder god for whom the warriors were sacrificed; and Teutates is the father of the tribes for whom the people were sacrificed. Like the bull, Esus, Taranis, and Teutates are all symbols of the warrior hero and the eternal return of this warrior in his many forms. Esus is the warrior hero cutting the branch of the tree of life to return in a new and stronger form, and Taranis holds the wheel, a symbol of solar power and continual renewal. Teutates is the god who immerses the warriors in the cauldron of eternal life. Acting as a triune of regeneration, they are much like the bulls who return in many forms in the myths and legends of the *Tain*.

The Gundestrup Cauldron accurately depicts the roles of these gods on each of its plates. Three bulls are sacrificed by three warriors on one plate; a small figure holds the wheel next to the head of Taranis on another; Teutates immerses warriors in The Cauldron of Revivification on another; and several gods of sacrifice, any one of them possibly representing Esus are on other plates. The bottom plate of the sacred cauldron depicts the killing of the bull and, most interestingly, a goddess is represented on the cauldron with birds on either side above her head and with rosettes, elephants and griffins (Kaul 103). Although the iconographic analysis of the cauldron suggests a medley of cultures with Thracian, Nordic, and Celtic influences, the underlying theme and meaning of the cauldron is one of sacrifice and renewal represented in the sacrifice of the bull and warriors who give their gift of strength through their eternal return.

The cauldron, the grail and the holy cup of life are potent symbols of the regeneration of the life forces present in the spring. These sacred vessels emphasize the importance of sacrifice and renewal. In the rituals and holy nights of spring, the third lunar cycle, the month of Ogronios on The Sequani Calendar, contains the Celtic cross-quarter day of Belteine, a celebration honoring the rites of regeneration, renewal, and resurrection. When the sacred marriage has been consummated, the bull sacrificed for the people and the abundance of crops, and the prayers for water answered, the third lunar cycle of spring welcomes the spring rains and the plant growth. The wheel of the year turns to a new season and the constellations change their pattern. The first star of the Summer Triangle, Vega, appears on the eastern horizon as the primary star of Ogronios and the spring comes to an end.

The rites of Belteine are a culmination of the sacred events of the spring. As a holiday for the people, Belteine includes all the elements of the celebrations of spring. The idea of purification and cleansing is represented by the running of cattle and bulls between two need-fires in order that they may be cleansed from vermin and ready to pasture in spring. The sacred marriage is represented in the Queen and the King of Belteine, a young couple who lead the other couples in merriment and welcome the honeymoon of the season. The sacrifice of The Dying God is represented in the eating of an oatmeal cake where the purposefully burnt section is arbitrarily chosen by someone who must leap the fire as the scapegoat and the symbolic sacrifice. And Resurrection is represented in the spiral and circle dances around the Bile or Sacred Tree that is erected as the center of tribal cohesiveness. For the people, Belteine represents a renewal of the vegetative spirit where sacrifice and prayer are conducted at the foot of the celestial column and direct communication with the spiritual planes is accessed. The chief Druid himself would be present at the rites of this auspicious occasion "wearing a bull's hide and a white speckled bird's headdress with fluttering wings" (Ross 83).

Although Belenos, the sun-god who represents the beginning of the season of light is the name-sake for the holiday of Belteine, the ancient festival itself is a culmination of the celebrations of the celestial spring evident in the sun, the stars and their constellations, and the moon in the night sky. The renewal and resurrection of the vegetation deities and their parallels in the night sky are the time-keepers of the eternal return of the spring and the abundance of the water of life. The rites of spring celebrated in the celestial realms have been with us since Neolithic times. Ogotemmeli, the high priest of the Dogon, a living Neolithic tribe in Northern Africa, reveals that the Golden Ram of the spring returns to his people each year and as he sinks into the Bananga marsh south of the Lower Ogol, the old man says that "He plunges among the water-lilies, crying: 'The water is mine! The water is mine!' "

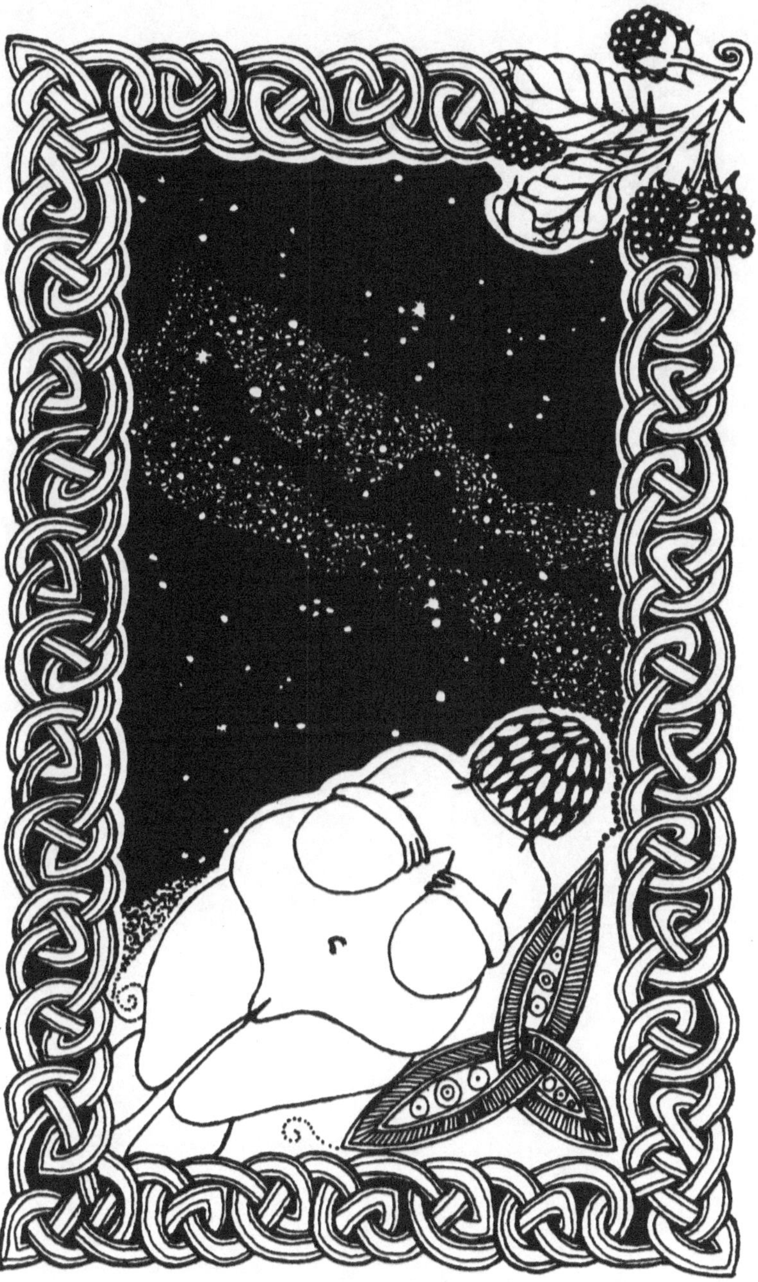

CHAPTER FIVE

Time-Keeper, Earth-Keeper, Star-Keeper

The lush verdure of the warm months of the summer solstice and the harvest of the berries, the vegetables, and the herbs bring the sky and the earth into harmony. The time-keepers take careful measure of the abundance and predict the coming year through the harvests of the earth. In the three lunar cycles of the summer months, the stars guided the ancient time-keepers through this season with precision. Like mirrors of the promise of immortal nourishment, they lull us into the belief of an eternal harvest, a cornucopia of abundance and a life of ease beyond this earthly summer. In a word, the summer of earth is the summer of eternal bliss echoed in heavens. In the myths of eternal paradise, the garden on the earth and the garden in the sky become one.

The paradise of the garden harvest is peopled with ancient and timeless symbols of the release of the energy of life and the need to protect and nourish that life. Prophecy of the coming year based on the harvest is ultimately important and renewal or replacement of the old ways is part of that process. The management of the earth's forces is vital on all levels. On a pragmatic level, the garden must be blessed and the fruits, vegetables and especially the herbs are used or stored for their year-long use. Among the herbs harvested in the summer, mistletoe, the golden bough, the elixir of medicines sacred to many ancient cultures is carefully tended. On an erotic level, the passion of life is released in the summer months and it too must be carefully managed as a resource. To the ancient peoples, the summer was the time to harvest the spirit as well. Eternal bliss is captured in the stars, the moon, and the sun as the stories of the heavens unfold the nature of the inexhaustible, great nourisher.

To the Neolithic peoples, the gracious mother pouring forth her riches in the creative fullness of the summer is represented as the mother of all living things. As bestower and helper, she is the force of full productivity in a beatific state (Neumann 15). Her iconography is a symbol group consisting of three basic elements where each is represented simultaneously or separately. The primordial waters necessary for the growth and harvest of the earth are one element, the earth serpent unleashing its energy is another, and the birds of prophecy, purification and ascendance are the third. Three is her sacred number and the summer constellations are her celestial emblems. Marija Gimbutas calls her the "Mistresses of the Waters: The Bird and Snake Goddess" (*Old Europe* 112).

The three aspects of the Great Mother Goddess of the summer months are represented in the heavens first, as the Milky Way, second, as Draco and The Serpent Bearer and the Serpent, and third, as the Summer Triangle. The Milky Way is the great river in the sky that represents the primordial waters of eternal life. In the summer months, the Milky Way divides the sky in half stretching from east to west at twilight and north to south as night unfolds signifying the absolute peak of the summer light. Light and water are ready to give their energies to the earth. The earth's powers are represented by the celestial serpents in the constellations of Draco, the "Old Serpent," and The Serpent Bearer and the Serpent. Releasing the serpent's energy is the uncoiling of the spiral of the Winter Solstice and the freeing of the serpent from the granary. The Summer Triangle consists of the constellations of Lyra, Cygnus, and Aquila which are the Vulture, the Swan and the Eagle, respectively. Each of these birds is the vehicle for purification, transformation and ascendance of the spirit. They are symbols of the air.

Through the elements of water, light, earth and air, the Great Mother Goddess brings eternal bliss to the people. The heavenly bodies are her representatives that urge the people to travel in the company of the spirits across the sky and her temples are stairways of discovery in the summer nights. The ancient peoples built monuments to honor the Great Mother Goddess across the world. In Cambodia, the impressive Angkor Wat temple-complex of the snake deity of Indo-China has a central causeway that dates to the Megalithic Era, and the foundations of the temple are patterned on the constellation of Draco (Hancock 126). Neolithic monuments that mark the Summer Solstice in

Europe and the British Isles are numerous. In Ireland, there are five mounds aligned to the Summer Solstice sunrise at Loughcrew and three mounds aligned to the Summer Solstice sunset in the Boyne Valley. Sels Kilgreen in County Tyrone reveals spiral patterns illuminated by the midsummer sun similar to those at Loughcrew (Brennan 55-87).

The spirals on the monuments, like the spirals at the Winter Solstice, are accurate representations of the forces of the earth seen as the snake coiled in the winter and sinuously uncoiling in the summer. The sun is seen splitting these patterns or their design meanders across magnificent stone walls that line the temples almost as if the energies of creation have come to life. Intricate patterns of uncoiled spirals resembling waves of eternal life are found on the walls of the chambered mounds of Gavrinis in Brittany, France. Here, the moon as well as the sun marks the advent of the Summer Solstice as the waves of creation breathe with the moon's celestial light. The shadow of the sun is also used by the ancients as a Summer Solstice designator at Avebury and Castlerigg in England where the sun leaves a shadow path, as long as two miles in the case of the latter. Finally, at Stonehenge, the sun rises over the Heel Stone at the Summer Solstice bringing crowds to celebrate the sky meeting the landscape (Cornelius 118-132).

The merging of the power of the Summer Solstice sun, moon and stars magnified by the symbol of the spiral and its uncoiling are all evident in the art and artifacts of the Neolithic culture as the earth symbolism merges with the celestial in perfect harmony. Earth unfolds its beauty when it is touched by the light of the sun, the moon and the stars. In the Neolithic culture, this merging of forces is represented by the waving line representing the earth energy of the snake and the forces of creation. The same wave of energy present on the monuments reveals itself in the art of the culture. Gimbutas sees this wave as the life-giving forces of the snake and the earth associated with "the upper waters." In their art, Gimbutas states that: "The eyes of the Bird or Snake Goddess even stare from the very center of the world- a sphere with a mythical water stream in the very centre" (*Old Europe* 112).

The mythical stream dividing the sky is most likely the Milky Way of the summer sky with the eyes of the birds of the Summer Triangle and the curves of the snakes of Draco and The Serpent Bearer surrounding this river in the sky. Like the images on the stones, the Neolithic art abstracts the fused forces of earth, light, water, and air into one picture of the creative forces at work in nature in this most abundant and fruitful time. Gimbutas comments that the images are at once separate and “a single divinity.” “Their functions are so intimately related that their separate treatment is impossible” (*Old Europe* 112). Her words “divine ambivalence” capture the essence of the Great Mother Goddess as the snake, bird, and water divinity of the summer sky.

The primordial waters represented as the Milky Way are indeed the most dramatic of the celestial representations of the goddess as it parts the sky in the same manner that the solstice light, the solstice shadow and the moon’s light parts and uncoils the spirals on the ancient monuments. According to Gimbutas, the water may be a symbol of the rituals of rain invocation by the ancients. The ancients appealed to the Great Mother Goddess for rain for their crops as she was the controller and releaser of the upper waters: “As a source of rain she was invoked as documented by the sculptures of nude rain-invoking women holding large basins, altar tables in the shape of animal-masked women holding vessels, bear-shaped cult vessels, and a persistently recurring decorative motif made up of rain-torrent parallel lines, zigzags, and dotted bands.” Rain-bearing and milk-giving motifs are interwoven where rain torrents and breasts appear on jars and basins used for rain invocation (*Old Europe* 145). The water of the Milky Way is indeed an accurate symbol of the life-giving source of divine moisture from the maternal skies.

The association of milk and water as nourishment needed for the growth of the summer garden does not disassociate itself from the snake or the bird aspects of the goddess. Indeed, the same artifacts from the Neolithic culture depicting the wave of the great life-giving force of the Milky Way are often snake or bird deities portrayed as mothers and nurses. The snake deities, represented by the constellations of Draco and The Serpent Bearer and the Serpent are powerful symbols common to ancient cultures. In Buddhist mythology, the snake, identified as Draco, the constellation that circumnavigates the celestial north pole, is the Naga serpent of creation uncoiling its body to release the forces of cosmic energy. This transdimensional creature crosses the sky and ground as a symbol of the sky and earth energy responsible for creation (Hancock 137).

In both Indic and Egyptian mythologies, the snake deity is able to create cosmic cycles of time by “churning the Milky Ocean” and pulling each epoch into being (Hancock 142). This is depicted in the sky where the constellation of Draco with Thuban, the star of the snake’s tail and the pole star of 4,000 years ago, anchors Draco around the celestial north to swirl around the earth initiating each new epoch of time. The ancient Greeks who depict the serpent as a force of creation in their myths, and most likely the Celts who see the serpent as the great god Nathrachain, as well as the Babylonians who depict the serpent as Tiamat, goddess of creation, were all conscious of the power of the snake as time-keeper and cosmic force of creation. Indeed, the serpent as a constellation seems to span the longitudes of the globe in each of their sectors, which, according to Graham Hancock, are marked by ancient temples such as Ankgor Wat, the Great Pyramid, and the Sphinx (182).

The Serpent Bearer and the Serpent are south of Draco in the summer sky and are related thematically by the obvious use of the snake imagery and, more importantly, as time-keepers of the cosmos. The Serpent Bearer and the Serpent known as Ophiuchus, Serpens Caput and Serpens Cauda to the ancient Romans, is a symbol of the Great Mother Goddess. The constellations depict a celestial figure who holds a huge serpent high in the sky as an emblem of power and as a celebration of the release of the creative forces. She is mistress both of the year and of the cycles of time as she initiates the fruition of the eternal paradise of the summer season. In Minoan culture, she is the snake-goddess of the palace of Knossos and in Greek mythology, she is the year as Hera. She is the power of the Great Mother Goddess displaying the uncoiled serpent as guardian, healer, and giver of fertility and wealth to the garden of the earth.

The Serpent Bearer and the Serpent as well as Draco are time-keepers in the sky, and in the calendars of the ancient Greeks and Celts, they appear in the myths of each season of the year. In the fall, the serpent is carried by Hermes on the caduceus and by Cernunnos in his hand as a symbol of the ability to master the creative energies of the serpent by subduing him and conserving his strength for the coming winter months. In the Skirophoria in Skirophorion or capella, the eleventh lunar cycle of the year, the serpent is blessed and placed in the granary to protect the harvest seeds. Cakes in the form of serpents are eaten as a symbol that he is hibernating and safely buried for the year. In the spring, the serpent emerges from the granary greeting the initiates of the Lesser Mysteries and decorating the collars of the dancers at the temple of Dionysus. Now, in the summer months, the serpent is held in high esteem by the Great Mother Goddess as her full power flourishes.

In the month of Cutios on The Sequani Calendar, the full moon passes through the foot of The Serpent Bearer as if to signify that this is the time of the emergence of the full powers of the Great Mother

Goddess. Her task is to protect and nourish the garden and her powers are immense. In the Neolithic culture, her role is central to the mythology and her serpent is the curator of that power. A living representative of the Neolithic culture, Ogotemmeli comments that the snake has "an enormous task to perform; as the mainspring of germination, the protector of placentas, the purveyor of vital force, he labours day by day with bursts of energy at critical times" (Griaule 118). The snake's connection to the maternal forces is evident and its task is of vital importance to the culture.

Gimbutas states that the snake, and its abstracted derivative, the spiral, are the dominant motifs of the art of Old Europe throughout the Neolithic (6500-3500 BCE) and Chalcolithic (3500-3000 BCE) periods. Snake-spirals were used as the basis of the entire ornamental composition when the art style reached its peak in c. 5,000 BCE. According to Gimbutas: "The mysterious dynamism of the snake, its extraordinary vitality and periodic rejuvenation, must have provoked a powerful emotional response in the Neolithic agriculturists, and the snake was consequently mythologized, attributed with a power that can move the entire cosmos" (*Old Europe* 94). The organization of the motifs, continues Gimbutas, is "genuinely cosmogonic" and associated with "the belts of the upper skies" and with belts of the earth characterized by plant motifs. To Gimbutas, these motifs are symbolic of the glorification of the dynamic forces of nature and the regenerating powers of the snake as a vehicle of immortality. Some of the vases depict a gigantic snake winding or stretching over the whole universe, "over the sun or moon, stars and rain torrents" (95). The snakes of the Neolithic art find their parallels in the constellations of the summer skies.

According to Gimbutas, when the Old European civilization reached its peak, "there emerged a sophisticated image of the Bird and Snake Goddess." Here, the image of the primordial waters, the snake and the bird merge into a figure often wearing a mask or with elaborate dress, medallions round the neck, and headgear. One late Vinča figurine of the goddess depicts a woman wearing a detailed dress and a mask of a water-bird. Others have spiraling snakes engraved on their necks or the upper part of the body: "Snake spirals, meanders, dotted bands or parallel horizontal painted bands often decorate the costume of the standing or seated Lady Snake" (*Old Europe* 136-37). The "chic coiffure," peaked caps, and spiral or snake coils on the goddesses' heads as well as their decorative costumes prepare the way for the priestesses of the powerful snake and bird goddess that dominated the palaces of Knossos in the cult of the Great Mother Goddess.

The bird epiphanies of the goddess often took the form of a duck, goose, swan and other water-birds as well as birds of prey. According to Gimbutas: "At all periods and in all regions of Old Europe one, three or six holes or notches were frequently cut into the throat of the Snake or Bird Goddess." Her associations with the number three and its duplication "possessed a certain symbolic significance where the goddess was concerned" (*Old Europe* 136). In the summer sky, the bird attributes of the Great Mother Goddess are represented by the constellations of the Summer Triangle. The Summer Triangle is fully risen and in the center of the summer sky in the sixth lunar cycle of the year or in the month of Giamonios on The Sequani Calendar. Three stars of primary magnitude, Vega, Deneb and Altair mark the three points of this isosceles triangle that dominates the summer sky; it is perhaps the most striking feature of the night and the most easily recognizable to the naked eye.

All three stars of the triangle are represented as birds of the goddess. Vega, which rises in the late spring, is a primary star in the constellation of Lyra, the Vulture or the Lyre. Deneb, which appears as the primary star of the month of Cutios, is in the constellation of Cygnus, the Swan, and Altair, the primary star of the month of Giamonios, is in the constellation of Aquila, the Eagle. At the peak of the Summer Solstice, the triangle is complete and magnificent as it straddles the Milky Way directly northeast of The Serpent Bearer. The rising of the birds of the Summer Triangle are symbolic of the transformation and ascendance possible through contact with the goddess. The spirit which is bound for the sky is represented in the Neolithic culture in the company of birds. As vehicles of the soul to its immortal destiny, the birds guided the ancients to the heights of the sky.

Birds of prey associated with the goddess, such as those depicted at the "Vulture Shrine" at Çatal Hüyük, swoop down on headless bodies to devour their flesh in a symbolic rite of purification. The eyes of the owl are also depicted on bare bones as symbols of death. The role of the birds of prey in the Neolithic religion is to remove the flesh from the bones of the bodies of the dead which are exposed to the elements for symbolic cleansing and purification. When the bones were clean, the dead were buried. According to Gimbutas, two raptors, the vulture and the owl, that predominate Old European mortuary symbolism both brought death and renewed life. Often, as in the case of the Lepenski

Vir site for the dead, the funerary artifacts in the Neolithic enclosures indicate complex excarnation rites (*Living Goddesses* 19-21; 109).

The celestial representation of the birds of prey, the constellation of the Vulture, is the first constellation of the Summer Triangle to appear in the summer sky. Like the rites of excarnation, it is symbolic of the cleansing and preparing of the spirit for the afterlife and rebirth. The second constellation, Cygnus the Swan, is the vehicle for parting the veil between this world and the Otherworld. In the Winter Solstice, the great goddess as the Swan parts the year and brings in the new year light symbolic of eternal light and life. In the summer nights, she is headed south bringing the remaining light and the waning rays after the Summer Solstice with her in her southern migration. Likewise, the swans and geese begin to fly south in the full moon of Giamonios. Her long neck stretches across the sky seemingly to transport souls to immortal worlds of light from the impending darkness. The water-bird, as a symbol of transport between two spheres, water and air, is most appropriate to carry souls and part the thin veil that separates this world from the world of the spirits.

The third bird to culminate the appearance of the Summer Triangle in the night sky is the Eagle as the constellation of Aquila. The eagle, able to reach the greatest heights, is a symbol of transcendence. In astronomy, the constellation of the eagle contains the primary star, Altair appropriately named "the Rising One" (Staal 179). The "Tomb of the Eagles" is a Megalithic burial site in the Orkney Islands of Scotland where excavators have discovered hundreds of bones entombed with the bones of thirty-five sea eagles (Gimbutas *Living Goddesses* 66). As in the case of the mythical Phoenix, the eagle is able to reach the heights of the universe and rise out of the ashes of death in a new form. Regeneration and rebirth through ascendance characterize the eagle and his strength.

The birds of the Summer Triangle act as stages in the purification, transformation, and ascendance of the soul. Their relation to the other aspects of the Great Mother Goddess is complex and inseparable from the creative energies of the waters and the snake. In the summer world of the ancient agriculturists, the union of the forces of creation of the earth and sky represented in the snake and the water are combined with the transformation of the soul. The season represents the life and death process where the creative life forces dominate. The underlying knowledge of the summer is that it contains both the peak of abundant plant and animal growth and the waning light after the solstice, a symbol of death. The solstice is the end of the light or the death of the light half of the year seen as the death of the sun. The hint of death-in-life represented by the birds of the Summer Triangle is not an image of futility, but an inspiration of renewal and rebirth at the center of the process of life.

In Minoan and Minoan-Mycenaean art, the presence of the Great Mother Goddess of the Neolithic era flourished. Gimbutas states that their art "abounds in birds, snakes and women with wings or with snakes crawling up their arms or on top of their heads, the epiphanies and the anthropomorphic images of the goddess inherited from the Old European pantheon." In Early Minoan art, the snake and bird goddess is depicted on cult vases, dishes and altar tables. At Phaistos, she is a beaked lady with snake curls on her head and snakes for arms. She is often associated with the meanders and water lines of the goddess as Mistress of the Waters. In the Palatial Period of the culture, the snake cult of the palace of Knossos central recipient is called the "Snake Goddess." In several exquisite figurines she is bare-breasted in a decorative costume and headdress and she is holding snakes at arm's length above her shoulders (*Old Europe* 146). Her stance evokes images of The Serpent Bearer and the Serpent in the summer night sky.

Birds are also associated with the snake goddesses from the shrines at Knossos, Gazi, and Gournia in Crete where they are perched on the head of the goddess, attached to her elbows by their tails, or found in groups of three or six with the goddess figurines. Again, the number three is clearly a number associated with the goddess and her birds. The birds portrayed on vases were exclusively diver or water birds such as geese, ducks, and gulls (147). Terracotta birds like the figurines of the goddess were discovered in the shrines of the palaces and the homes indicating the bird as part of the domestic house cult of the goddess. According to Martin P. Nilsson, they were rarely found in the votive deposits of the cave and the open-air sanctuaries in comparison with the masses of other animal figurines. Nilsson states: "It appears that the birds are not votive offerings, but representations of the epiphany of a deity" (335). The deity Nilsson refers to is most likely the snake and bird goddess of the Minoans descended from the Great Mother Goddess of the summer months.

In the Central Palace Sanctuary at Knossos, evidence of a house cult of the bird and snake goddess is abundant. Curious tube-shaped vessels and colanders or sieves with snakes on them are part of the cult apparatus suggested by Nilsson to be used in the house cult (88-92). Nilsson affirms the fact that the palace artifacts are evidence that the snake and bird goddess of the Minoan culture was indeed a cult centered in the palaces and the homes of the people. According to Nilsson, the house snake is well known in European folk-lore where even in modern Greece, Albania, Lithuania and Serbia the snake appears as "the genius of the house." The snake is fed with libations of milk and welcomed as an honored deity in the homes and gardens of the people. Such a snake in Herzegovina and the Serbian lands may be called the domachitsa or "house mother." Snakes protected the children and cattle from diseases spread by vermin and were therefore welcome to live in the walls or foundations of the house (325).

The associations of the snake and bird goddess with the house cult strengthens in the art of the Mycenaean culture. The pragmatic value of the snake in the home and garden especially in the summer months is seen as a vehicle for the goddess' role of protector and guardian to emerge. In Mycenaean art, the goddess is depicted seated on a throne, decorated with the meanders of the primordial waters and frequently wearing a crown. The coiffures and headpieces of the goddess of the Neolithic peoples have now become her crown. In the Baltic religion, the crown is a symbol of wisdom and wealth. Gimbutas states: "The crown enables one to know all, to see hidden treasures, and to understand the language of animals (*Living Goddesses* 204). Crowned and on her throne, the goddess possesses the knowledge of the abundant earth ready to rule and protect her home.

In Classical Greek mythology there is ample evidence of the presence of the Great Mother Goddess of the Neolithic, Minoan, and Mycenaean eras. Although influenced by the Indo-European migrations, the images of the water, snake, and bird goddess remain distinguishable in the mythology of the Greeks. Like the goddesses of the other seasons, the characteristics of the goddess of the summer months are represented by separate goddesses thus dispersing her power and influence. These goddesses are often married to the solar deities or sky gods and demoted in position. Negative aspects of the goddess emerge and the mythology as history reflects an era of violence and submission. Fortunately, many aspects of the goddess of previous eras are depicted in the written mythology that were indiscernible in the artifacts of the Neolithic, Minoan, and Mycenaean civilizations. The connection of the goddess with prophecy, her role as time-keeper, her maternal attributes, and her position as guardian of the house and garden of eternal bliss are given literary characteristics in the mythology. Finally, in the heroic myths of the ancient Greeks, the Indo-European concept of the hero as slayer of the serpent and the transference of the powers of the goddess to the hero are as evident in the stories of the summer goddess as they are in the other seasons.

The primordial waters associated with the Great Mother Goddess are represented in Greek mythology by a group of fifty goddesses or priestesses called the Danaids. Their name is linked to a Neolithic tribe which inhabited an area near a watery river or stream, thus the word "**DeHanu*" or **Donu*, is cognate with the word for river in Proto-Indo-European and Indo-European languages. According to Dexter: "This water became personified as a feminine deity, and the goddess gave her name to the tribe. The tribe was probably not Proto-Indo-European but one with which the Proto-Indo-Europeans came in contact early on" ("Reflections" 58). The tribe, or people of Danu, were subdued by the Indo-Europeans and the watery "monster" that was once the goddess is punished or slain. In the Greek myth of the Danaids, the water goddesses are punished because they murder the fifty grandsons of King Belus, a figure associated with the ancient solar deities.

Aegyptus, the son of Belus, has already conquered the country of Melampodes and named it Egyptus after himself. The arranged marriage with the daughters of Danaus, the Danaids, would insure his further military success in Libya. Danaus and his daughters not wanting to succumb to Aegyptus, flee to Argos where the king of Argos resigns his throne willingly because Danaus and his daughters are able to bring water to Argos which has suffered a drought. Danaus becomes such a powerful ruler that the Pelasgians of Greece now call themselves the Danaans. He builds the citadel of Argos and his daughters bring the mysteries of Demeter to the Pelasgian women as well as discovering three sacred wells, the eternal source of the sacred river of Lerna from which each year Dionysus is revealed (Graves, Vol. I, 201).

Clearly, the Danaids are important figures in Greek mythology. Even though the myth states that the Danaids came from Libya, which may or may not have historical parallels, the symbolic message of the myth is that a distant people came to Greece to instruct them in the ways of the goddess and bring them eternal life through the worship of the sacred source of life: water. The Danaids give the Danaans the rituals of Demeter, the source of life as the sacred waters of Lerna as Dionysus, and they instruct the people in the art of sinking wells. When Aegyptus pursues them to Argos and forces them to marry his sons, they recapitulate by murdering their husbands on their wedding night. The Danaids are absolved of their crimes by Zeus in their purification in the Lernaean Lake; however, they are condemned by the Judges of the Dead to carry water in jars perforated like sieves.

Although the Danaid's punishment is typical of the new sky god and his judges to institute change on a colony of goddesses or priestesses, the Danaids' punishment incorporates an older tradition into a new mythology. The sieves or colanders used by the rain-invoking women in Neolithic culture and the cult apparatus decorated by snakes in the Minoan culture are evidence that the priestesses of the goddess used water utensils in the cult of the bird and snake goddess in previous eras. Most likely, the sieve was used as a rain-making device where water was sprinkled on the fields to invoke the goddess for rain for the summer crops. Indeed, the sieve, or leaking pot, remains a distinguished mark of the wise woman in the writings of Philostratus many centuries after the abolition of the Danaid colleges (Graves Vol. I, 205).

In Greek mythology, the powers of the primordial waters and the snake are combined with the powers of the earth goddesses. The snake is represented as an entity imbued with the powers of creation. In the Pelasgian creation myth of the Greeks, the great serpent Ophion coils about the limbs of the goddess Eurynome and couples with her. Like the great serpent of the sky, Draco, who coils about the celestial north pole, Ophion too comes from the north surrounding the goddess with his powers. In the myth of Boreas, the north wind, Boreas carries away Oreithyia, the daughter of Erechtheus, King of Athens and ravishes her. Boreas has both serpent tails and wings, and again is associated with the north. The Athenians built a temple to him on the banks of the river Ilissus and call him the brother-in-law of the Athenians (Graves, Vol. I, 170). Ophion and Boreas, great serpents of Greek mythology, like the serpents of many mythologies, play a vital role in the original creation and the creation of cities in the Greek myths.

As center of the universe and creation, the serpent also has prophetic powers in the mythology of the Greeks. At the oracular shrine of Delphi, the Python guards the omphalos, or naval, of the Greek world. Delphi was once owned by Mother Earth who resigned her rights to the Titaness Themis, mother of the Seasons and the Three Fates. Her shrine is guarded by Python, a huge female serpent, who is eventually killed by Apollo when he seizes the temple. The Pythoness, a priestess for the temple, still prophesizes for Apollo by receiving messages from the snake. The Pythoness receives messages from a secret well and pronounces an oracle in verse. To receive the oracle, the inquirer is given a message from a ghost in serpent form (Graves, Vol. I, 180). Perhaps, the idea of the snake with oracular powers came from the ability of the priestess to determine the outcome of the year by the water level in the wells and crop production of the summer months as time-keeper for the people.

The powers of prophecy are given by the snakes to the prophets. Although the snake is a representative of one aspect of the goddess, the images of water, as seen in the above mentioned myths of the snake as prophet, and its association with birds still occur in the mythology. In the myth of Melampus, the role of both snakes and birds play a part in giving their gifts of prophecy to mortals. Melampus saves a brood of young snakes, and when they are fully grown, they lick his ears when he falls asleep giving him the powers of prophecy. Immediately, Melampus is able to understand what the birds are saying and he is able to prophesize the future to men. The prophet is seen as a man who must be able to meld the knowledge of two realms associated with the goddess in order to fully understand the powers and forces of creation. Dexter notes the continuing interrelationship of the birds and the snakes in Classical mythology: "while the snakes gave the gift of prophecy to Melampus, it was the conversation of the birds which formed the material for the prophecies (*Whence the Goddesses* 8).

The Olympian goddesses of classical Greek mythology present the same dichotomy as the water and snake deities. On the one hand, they represent an aspect of the Great Mother Goddess and are not the comprehensive entity of earlier civilizations, and on the other hand, the elements associated with the goddess are inseparable and often represented as details in the myths about her. In the cases of Hera and Athena, several aspects of the goddess are represented, but she is still divided into many distinct figures in the mythology. Both are the bird and snake goddess with different attributes. Hera represents the goddess as keeper of the seasons of the year, mistress of the house cult, and guardian of the garden. She is associated with snakes and birds, but is predominantly the guardian of the home as a maternal figure. Athena, although a guardian and protector of the city of Athens, is closely associated with the birds of death, the vulture and the owl, the politics of war.

In the Olympian pantheon, Hera or **yera*, is married to the sky-god, Zeus, as goddess of the year, thus her status changes and she is submissive to his will, perhaps as a symbolic act of the calendar changing from the calendar of the goddess to the new solar calendar. In the Minoan Bronze Age, Hera's and Zeus' names as well as Dione's and Zeus' names appear on the Linear B tablets side by side, and in pictorial representations, Hera is shown occupying the throne with Zeus standing at her side (Gimbutas *Old Europe* 149). Hera's Minoan heritage is evident, yet her status has changed. Every year Hera is able to maintain some of her power by returning to her virgin state before she again is forced to marry Zeus. In the myths of the ancient Greeks, Hera is still associated with the time-keepers, but her status has been demoted by the appearance of the powerful sky-gods of the Olympian pantheon. Her virginity is intact, but her absolute power has diminished.

Fortunately, she is still the keeper of the Garden of the Hesperides, a mythical paradise of lush and abundant greenery, a symbol of the everlasting land of summer which Mother Earth has given to her. In the Garden of the Hesperides, Ladon, a great and powerful serpent guards the golden apples, symbols of immortality. Still connected to The Serpent Bearer and the Serpent as guardian of the eternal garden of bliss, Hera also retains the powers of prophecy associated with the serpent. Hera was equal to Zeus in one power alone: she could still bestow the gift of prophecy on any man or beast she pleased (Graves, Vol. I, 53). In this way, she may still be seen as mistress of the cosmos and ruler of man's fate. The prophets are subject to her will and receive her oracles from the ancient serpent. The apples of the garden are Hera's gift to the heroes. They are a symbol not only of eternal life, but of the abundance the summer garden has to offer.

From her finely crafted throne in her sanctuary, Hera retains her ancient power as bestower of the gift of prophecy upon those she wished to favor, human or animal. According to Dexter: "It would seem that the concomitant function among such goddesses was that of the priestess, represented as the ability to prophecy and to have foreknowledge of all things" (*Whence the Goddesses* 122). Her temples and sanctuaries in Corinth, Argos, and on the islands of Samos and Lesbos are built to reflect her associations with the snake as prophet of the priestess and guardian of the house and garden. She is portrayed in the artifacts from her temples with hair curling like snakes and with a snake winding or zigzagging vertically in the middle of her skirt. Votive offerings in her temples include terracotta snakes and on a Boeotian bell-shaped idol, decorative birds hold snakes in their beaks. (Gimbutas *Old Europe* 149).

Hera's sanctuaries were situated in valleys at the estuaries of rivers, near the sea, and surrounded by pastureland, and Hera is often referred to as "cow-eyed" or "ox-eyed." Votive offerings from the sanctuaries include horned animal figurines, calves, and anthropomorphic idols with large eyes, decorated with spirals and meanders. Shrine models are decorated with similar designs. Known as the "Giver of All" and "the Noble One," Hera is depicted as a maternal figure (Gimbutas *Old Europe* 150). Her association with cows and calves, rich lands, and pure waters that make pure milk evoke images of the Great Mother Goddess of the Milky Way, the river of the summer night sky. Despite her constant bickering and attempts at jealous reprisal at Zeus' choice of mistresses in the man-made myths, she is still represented as a powerful female figure among the goddesses and gods of Olympia. A symbol of her power as wife and mother is the fact that she gives birth parthenogenetically to Hebe, the cupbearer of Olympus represented by the snake in the form of the constellation of the Hydra, and her two brothers, Hephaestus, the smith, and Ares, the god of war.

The myths of Athena's birth illustrate her affiliation with the bird and snake goddesses as mistresses of the waters. In the Pelasgian myth of Athena's birth and childhood, Athena is born in Libya, the homeland of the water-deities, the Danaids, beside Lake Tritonis. When engaged in friendly combat with her playmate, Pallas, she accidentally kills Pallas and sets her name before her own. Athena then comes to Greece by way of Crete to live near the Boeotian River Triton (Graves, Vol. I, 44). Like the myth of the Danaids, the myth of Athena coming to Greece from Libya may parallel history, more importantly, the myth records an accurate metaphor for the beginnings of the cult of the goddess in Greece affirming her association with the ancient water deities in order to preserve an important aspect of her character depicted by former cultures.

The fact that Athena takes the name of Pallas illustrates her affinity to her past. Athena is also called by the cult-name Tritogenia or "born of Trito," which might be a derivative of the bodies of water referred to in the myths. A second myth concerning Athena's birth, also mentions the shores of Lake Triton (Graves, Vol. I, 46). Here, Zeus gives birth to Athena himself, as a symbolic gesture of the patriarchy creating a new identity for the goddess of the previous eras. After impregnating, Metis the goddess of wisdom and counsel, Zeus receives an oracle from Mother Earth that his heir will depose him. He swallows Metis and gives birth to the fully-armed Athena when his head splits open as he walks on the shores by Lake Triton. Dexter describes Athena as a patriarchal deity "par excellence" where the birthing by a god of a "pre-Greek goddess could be assimilated into the patriarchal Indo-European pantheon" (*Whence the Goddesses* 119). As the daughter of Metis, Athena inherits the ability to counsel in war and, as a fully armed warrior, she is able to protect her homeland.

Despite these changes made to the image of the goddess, her affiliation with former images is still intact. As a representative of the snake attribute of the goddess, Athena is depicted with the snake at her feet, concealed under her shield, or by her side "in equal height or majesty." According to Gimbutas and Nilsson, the Indo-European and Oriental influences on the goddess were due to the protectress of the snake and house cult becoming the protectress of a city naturally when the city of Athenai or Athens, named after the goddess, became a city at war (Gimbutas *Old Europe* 149). The naming of the city involves a dispute between Poseidon and Athena where Athena wins because she takes possession of the city in a gentle manner by planting its first olive tree beside Posiedon's well.

Athena's gift of the olive tree is depicted on a well-known vase in Kertsch that represents the contest of Athena and Poseidon and is believed to be a copy of the central part of the western pediment of the Parthenon (Nilsson 497). Here, Athena is represented as giving the Athenians a version of the garden of the Great Mother Goddess complete with a tree and a guardian snake. In Greek mythology, Athena also gives the gifts of weaving and spinning, the earthenware pot, the plough, the rake, the ox-yoke, the horse-bridle, all symbols of domestic culture and the house-cult. Her tactics and strategy in warfare are to protect and guard the land by upholding the law through good judgements, counsel, and pacific means. Her role is symbolic of this ancient house-cult now in the form of protection of the city-state. Although she has many acropolis sites sacred to her, such as Argos and Sparta, the citizens of Athens look to her for protection and claim their heritage to her through mythology.

The myth that claims the Athenian's heritage to the goddess begins when Athena asks Hephaestus, the smith, to make her a set of arms, and he mistakenly believes she will pay him through love. He makes violent advances to her under those beliefs. As Athena tears herself away from the smith, he ejaculates above her knee and in disgust, Athena throws the sperm on the ground accidentally fertilizing Mother Earth. When Mother Earth refuses to accept the responsibility for the child, Athena takes it and raises it as her own. Erichthonius, the child born half-serpent and half-man, is hid in a sacred basket and when discovered accidentally by the women of the Athenian king, takes refuge in Athena's aegis. Athena rears him tenderly and many mistake her for his mother. Later, Erichthonius becomes King of Athens and the royal family call themselves Erechtheids; "they used to wear golden serpents as amulets and kept a sacred serpent in the Erechtheum" as a proud symbol of their lineage from the goddess (Graves, Vol. I, 97-99).

Athena's affiliation with the powers of the snake as protector of the house and city is represented by her aegis in mythology which originates on the palladia or small shield-bearing figures found in Minoan and Mycenaean art. According to Nilsson, the palladium of the house-goddess was hidden in a secret place in the interior or citadel of the temple as the symbol of the people's welfare and safety (500). These shield-bearing figures were protectors of the house and not warriors with combat shields. In the same way she protected the Mycenaean king and his needs, the goddess now protects the Greek Iron Age peoples. The aegis with its Gorgon mask is a symbol of that protection on the chest of the palladium of Athena.

Athena's affiliation with the bird goddesses is depicted in the artwork of the Iron Age Greeks. According to Gimbutas, Athena has shed the bird-form, but she is occasionally winged and the bird is her attribute: "She sometimes appears in the semblance of a sea eagle, a gull, a swallow, a vulture, or a dove, thereby perpetuating the Minoan-Mycenaean tradition." On a Corinthian aryballos, she is depicted as a diver-bird, and in the earliest known vase-painted representation of the rape of Cassandra a large human-headed bird stands behind the figure of Athena (*Old Europe* 148). Most often associated with the owl on the coins of the Athenians, Athena's bird attributes span all the birds represented in the Summer Triangle.

Athena's owl attribute may represent the "dark" or death aspect of the goddess (Dexter *Whence the Goddesses* 119). Athena's aspect of the vulture may also represent her dark nature and its earliest associations with the Neolithic excarnation rites. In the Iliad, Athena and Apollo alight in a tall oak in the form of vultures to watch the battle (Nilsson 491). Perhaps more than any other change made to the goddess, her association with deaths during war most accurately depicts the Indo-European influence on the Great Mother Goddess. Although Hera is "a first-function sovereign goddess" through her powers of prophecy and her sovereignty over the land, Athena is the more active goddess in martial affairs. According to Dexter, an Indo-European female figure in war bestowed power upon the warrior by giving him strategy, courage, and inspiration in battle. As strategist and endower of martial energy, Athena has an active role as protector of the warrior class (*Whence the Goddesses* 152-53; 157).

The Iron Age Greeks changed the images of the goddesses and gods in mythology as well as the images of the goddesses and gods in the sky in order to accommodate Indo-European influences and the changing ideology of the heroic age. The Serpent Bearer and the Serpent are named Ophiuchus and the constellation of Hercules, faint and above the head of The Serpent Bearer, is named after the Iron Age Greek hero. Although Ophiuchus is named after the gigantic serpent that Hercules kills for Omphale who purchases Hercules as her slave and later falls in love with him, the constellation is also affiliated with Asclepius by astronomers and in the mythology of the Greeks. Asclepius, the son of Apollo, learns the art of medicine from Apollo and Cheiron the Centaur, and is given sacred blood drops from Athena's serpent, the Medusa, that give him the powers to raise the dead; Erichthonius is given the same phials of blood. When Zeus hears of Asclepius' ability to raise the dead, he kills him but honors him by placing him in the sky with the serpent (Graves, Vol. I, 174-75).

In another version of the myth, Asclepius kills a snake and to his surprise another snake arrives and revives his companion with herbs. This is how Asclepius learns of the medicinal value of plants and is known as the god of medicine; the caduceus, his staff entwined with snakes, becomes the emblem of the medical profession (Chartrand 558). When Zeus hears that Asclepius has the knowledge of the elixir of life, he sends Aquila, the Eagle, to kill Asclepius with his thunderbolt: "By studying the movements of the constellations with a planisphere one can see how the Eagle stands threatening over Ophiuchus [Asclepius], as he slowly sinks below the horizon in the west on an Autumn night" (Staal 195). Thus, Asclepius dies in the mythology and in the sky.

Although the Great Mother Goddess is associated with the abundance of the garden and with the powers of the snake as prophet, it is not until the Iron Age mythology that some light is shed on the possibility of her role as medicinal healer and herbalist. The association of Ophiuchus and Asclepius with the snake, and more specifically with the medicinal powers of the snake, might have been powers attributed to the goddess of earlier eras. Hera's association with the golden apples as symbols of immortal life given to the heroes as their passport to paradise, and Athena's surname of Hygieia, or health, are clues that indicate possible connections, but the fact that the newly named constellation of Ophiuchus has a strong affiliation with medicine and herbs contributes to the theory of the Great Mother Goddess as herbalist and healer. Her role as herbalist and healer is further justified by the natural fact that herbs are harvested in the three full moons of the summer months.

The fact that Asclepius must be killed because he has been given the power of the snake is symbolic in the sense that it is leading up to what Dexter calls "The Fall of the Sacred Serpent" in mythology. According to Dexter, the "serpent is a power to be reckoned with, and consequently to be feared." "Therefore, in the mythologies of many post-Neolithic societies, serpent deities representing negative forces were pitted against young hero-deities. Often, one culture housed both positive and negative myths about the serpent" (*Whence the Goddesses* 9). In Greek mythology, Dexter notes that female monsters, like the Furies, the Gorgons, the Harpies, the Dirae and the Sirens were "dread creatures" descended from the snake and the bird goddess their potency held in check by the Greek and Roman cultures (180-81). They are challenges to the heroes of Greek mythology to overcome the forces of the goddess as monster and assume her powers.

Although many of the Greek heroes are responsible for the death of the serpent in Greek mythology, the most obvious assimilation of the powers of the snake and bird goddess and the closest association with Hera as the Great Mother Goddess are depicted in the myths of Hercules. Hercules conquers Hera by assuming the knowledge of the goddess. In the myths of Hercules' birth, Hera is tricked into nursing Hercules thus endowing him with the milk of heaven, symbolic of the primordial waters seen in the summer sky as the Milky Way. In the myth, when Hera realizes that Hercules, the son of Zeus and Alcmene, is nursing from her breast, she turns him away violently, thus forming the Milky Way from the milk that spurted from his mouth (Graves, Vol. II, 90). Inadvertently, Hercules is given the nourishment that will eventually immortalize him through the milk of the Great Mother Goddess.

In Hercules' conquering of the snakes, he assumes the guardian powers and strength of the goddess to protect the people. When Hera sends two "prodigious azure-scaled serpents" to kill him in his crib, Hercules strangles them one in either hand (Graves, Vol.II, 91). In the Twelve Labors imposed on Hercules by Hera and translated through the oracle of the Pythoness, Hercules must redeem himself for the murders of his children to receive his immortality. Hercules is presented with a shield with twelve serpents embossed on the front to help him in his labors. In his sixth labor, Hercules conquers the Stymphalian birds and three shrines are founded in Hera's honor because she is said to have retired in Stymphalus (Graves, Vol.II, 120). The conquering of the snakes and birds of the goddess is the first stage in Hercules' assuming the powers of the goddess.

The second stage is Hercules' journey into otherworlds of Greek mythology, the Garden of the Hesperides and Hades. In the Garden of the Hesperides, Hercules is sent to fetch the golden apples of the Hesperides by killing the hundred-headed serpent, Ladon, that guards the tree upon which the apples grow. When Hercules kills Ladon, Hera sets Ladon in the sky as the constellation of the Serpent (Graves, Vol.II, 146). Hercules is then sent to Hades to capture Cerberus, the three-headed dog that guards the gates of Hades. Symbolically and literally, Hercules' second stage of becoming a hero and receiving immortality as well as the powers of the goddess involves transgression of the otherworlds to conquer mortality. Hercules is successful at both tasks, thus he becomes the archetype of the Greek hero.

The apotheosis of Hercules is the final stage of his assuming the powers of the goddess. In the myth of his apotheosis, Hercules is burned on a pyre of oak and wild olive branches and "like a snake that has cast its slough, appeared in all the majesty of his divine father." His mortal part has been burned by the fire and his immortal part ascends to Olympus to be with the goddesses and gods eternally. He is adopted by Hera in a ceremony of rebirth and given her daughter, Hebe, in marriage (Graves, Vol.II, 203). The constellation of Hercules depicts his apotheosis in the sky. In the constellation, he is kneeling with a club over his head directly aimed at the head of the Serpent. He has conquered the serpent and possesses its powers eternally. As both suppliant and conqueror, Hercules has become the son of the goddess and the master of the serpent.

In Celtic mythology, the evolution of the Great Mother Goddess undergoes changes in both the images used to depict the goddess and in the thematic context of the tales. The goddess retains her associations with the maternal waters, the snake and the bird, but, like in the Greek mythology, she is often depicted as separate figures in the myths. She is associated with sovereignty and protection as well as healing, and eventually she gives these powers to the Iron Age heroes. Her powers of prophecy are still depicted, but they are more readily identifiable with birds as prophetic sources than the snake. Druidic divination rites and myths focus on the raven, the eagle, the swan, and the wren as sources of healing and augury in their ornithomancy. Moreover, the role of the goddess as a prophet of blood and death in warfare is more dominant in the myths of the Celts than in the myths of the Greeks. To the Celts, the goddess is a woman in an egalitarian position with Druidic priestesses representing her on the battlefield, in the Druidic orders, and in the home.

The primordial waters of the Great Mother Goddess in Celtic mythology are represented by Danu, the ancestress of the tribe. Like the Greeks, the Celts depict a migration of peoples, perhaps in the Bronze Age, from continental Europe to the British Isles. Again, whether these people, like the people of Danaus from Libya and Crete are an historical reality or not, is over-ridden by the fact that they are representative of migratory peoples that worshipped the river goddess Danu and brought their ideology and worship with them to the indigenous peoples of Britain in an effort to affirm the importance of the rivers and the sacred waters of life as a central focus of goddess worship. The people of Danu or the Tuatha Dé Danaan, like the Greek Danaans, settled the British Isles bringing their healing arts with them. In Wales, the goddess is Donu, Dana or Don, in Northern Britain, she is Verbeia, and in Ireland, she is Danu, Anu, Ana, Boann or Boand.

Although many of the sources of rivers and sacred springs in the British Isles are associated with water nymphs and goddesses, the river goddesses that are directly associated with the Great Mother Goddess are those with snake, bird, or water and milk giving aspects. Danu, a goddess of boundless energy, giving life and nourishment, wealth, and treasures is the most revered as the ancestress of her tribe. In Irish mythology, the Tuatha Dé Danaan give four gifts to the people: the Lia Fail, the Stone of Destiny symbolic of the powers of prophecy of the goddess, a spear and a sword, symbolic of the powers of protection of the goddess, and a cauldron, symbolic of the primordial waters and life-giving powers of the goddess. These gifts are given to the people of Danu as treasures representing the knowledge and abundance of the bird and snake goddess.

The Tuatha Dé Danaan also possess the powers of magic and druidry that affect all aspects of the people's lives from battle to the healing arts. Their ability to approach situations with mental powers of persuasion and their ability to move the waters, the rocks, the animals and the trees and all the natural forces is symbolic of their spiritual symbiosis with nature. In essence, they are the people of the earth who have mastered its energy and put it to their own use. Their knowledge of the earth and sky comes from their worship of the goddess as the source of that life. In the battles against the Fir Bolg, the Fomorians, and finally the Milesians who defeat them and take the island, the Tuatha Dé Danaan use their knowledge of the spirits of nature for political bargaining and for battle.

In the end of their struggle with the other tribes of Ireland, they are defeated at the battle of Tailtinn, but they do not die or disappear. They learn the arts of husbandry and farming from their enemies, when to plough, sow, and reap, and then they retreat to the mounds of the ancestors proclaiming peace and victory for themselves. They reside in the mounds and earthworks known as the Sídhe. As goddesses and gods of the Otherworld or People of the Sídhe, they co-exist occasionally making visits to this mortal world to give support or tantalize mortals with their music and magic (Heaney 3-21). Unlike the goddesses and gods of other religions, these pagan deities do not take on new names or forms, but co-exist with the goddesses and gods of future religions in a peaceful and yet always present form. Although they have many attributes and associations with other powers than their original name suggests, they honor their ancestress, the goddess Danu, through retention of her name as their own.

In Britain, a female deity on a water-jar cover may be a representative of the water aspect of the Great Mother Goddess. In the center of a group of figures, a female deity decorated with spirals and curved lines has a bird beak and she is seated in such a way as to suggest a squatting posture (Ross 278). Her identity is nameless, yet as "a patron of potters whose water-jar cover she adorns," she is comparable to the figurines of the Neolithic goddesses Gimbutas describes. In Northern Britain, the goddess is worshipped as Verbeia, or "She of the Cattle." As goddess of the River Wharfe, Verbeia is depicted with a large head, perhaps resembling that of a cow, a long pleated dress, and snakes in each hand. On a relief from Mavilly, Cote-d' Or, France, this same water goddess is also depicted with a long pleated dress holding snakes (Ross 279). Although an older version of the water, snake and bird goddess, it is clear her descendents are found in the British Isles.

In the mythology of the Celts, Boann or Boand and Verbeia are represented as the aspect of the river goddess as cow, similar to Hera in Greek mythology and Danu in Indic mythology. Known as "She Who Has White Cattle," Boann is a cow goddess of the Tuatha Dé Danaan who mates with the Dagda over the River Boyne thus giving it and the Brú na Bóinne patronage and her name. There is one myth that depicts her tragic death. As the wife of Nechtan, she is forbidden to visit the sacred well called the Segais Well, a source of inspiration and knowledge. Boann violates that geis and walks around the well in a left-hand circle. The well rises up and drowns her forming the River Boyne which is deemed sacred to her. In another version of the myth, Boann escapes and does not die (Ellis 41). Whether she escapes or dies, she is immortalized in the mythology as the goddess most closely associated with the waters of life as the milk-giver, nourisher, and mother goddess as cow.

Another aspect of the Great Mother Goddess, the snake as protector and guardian, is evident in the mythology of the Celts. According to Ross, ram-headed, crested and aquatic serpents were of considerable importance in the mythologies of both the European and insular Celts. The earliest evidence suggests a cult to the serpent god, Nathrachain, with the Stone of the Serpents dedicated to him in Scotland. Other serpents associated with rivers, wells and thermal waters often depicted with wings were found in Britain, Scotland and Wales (Ross 433-34). Like the serpents that Cernunnos holds and the serpent Conall Cernach has in his belt, the serpents that guard the rivers and wells are symbolic of the snake as protector and guardian of the treasures and wealth of the people. The relationship of the snake as guardian although associated with the gods still retains its associations with the goddess through the recurrent imagery of the waters, the wings of the bird, and the crest or crown as a symbol of the sovereignty and protection of the goddess.

In the mythology of the Celts, the snake is both friend and foe to the gods and, like its counter-part in Greek mythology, the gods must assume its powers and conquer it simultaneously. In the metrical Dindshenchas, Diancecht, the god of healing and the toreutic arts, must kill a great destructive serpent and assume its powers in much the same manner that Asclepius conquers the snakes and is given the knowledge of medicine. In the prose Dindshenchas, Diancecht must also slay the three snakes found in the heart of Mechi, the son of the Morrigan, burn their hearts and cast their ashes into the river (Ross 199). Diancecht of the Tuatha Dé Danaan through his association with medicine and the sacred serpent is a remnant of the goddess as healer and herbalist.

In Druid ritual, the aspect of the goddess as healer and herbalist is preserved in the rites of harvesting mistletoe at the Summer Solstice. Because of its unusual properties, such as having both flower and fruit on the same branch and being a cure-all and inhibitor of cancer and leukemia, mistletoe is the symbol of the powers of the healing arts in Celtic lore. Named the Golden Bough of elixirs, mistletoe is both sacred and sacrosanct to the Celts: "Not quite herb, not quite tree, beyond the limitations of classification, freed from the restrictions of convention, and resembling a constellation of stars suspended in midair from the bough of a sacred tree--such is the 'spirit' of this plant" (Hopman 101). According to Pliny, on the sixth day of the lunar cycle of the Summer Solstice, mistletoe is harvested by the

Druids (Hopman 103). The Golden Bough is cut and the most potent and powerful plant taken from the garden of the Great Mother Goddess.

Unusually enough, mistletoe is only spread by the droppings of birds, and it is the birds of the garden of the goddess that play a major role in the mythology of the Celts. Like the snakes, the birds are seen as guardians of the treasures of the people. In Celtic mythology, the garden of the Otherworld is a sacred island of immortality that contains the treasures of Thisworld, such as abundance of song, banqueting, and the fruits of the earth. Here, the pleasures of the earth are guarded by the magic birds of the goddess. According to Ross, "this is the idea of an otherworld goddess or goddesses concerned with sexual love and fertility, and possessed of magic birds by whose singing all sorrow is forgotten, pain healed, and even the dead are restored to life" (340). In the Welsh Mabinogi of Branwen, three sacred birds guard the Otherworld of Rhiannon and, in the Irish tales of Clidna, three spectacular multi-colored birds enchant the heroes in the goddess' Otherworld. In the sidhe of Dornbhuidhe, the birds also chant welcomes into her land of promise (337-340).

Like the birds of the Otherworld who often appear in Celtic mythology in a triskele, the birds of the raven-goddess appear in three forms as the Badv, Nemain, and the Morrigan. As descendents of the Great Mother Goddess, the raven-goddesses of Celtic mythology are associated with the house and land as sovereign, the powers of prophecy and the politics of war. These dark birds appearing as crows, vultures and ravens are represented in the summer sky by Vega in the constellation of the Vulture, the first star of the Summer Triangle rising in the east. According to Ross, they belong to "a wider European solar phase" and often appear with swans and other water birds (311). Their solar associations and their appearance with water birds in the iconography of the Celts parallels their appearance at the Summer Solstice and their association with Deneb, the swan of the Summer Triangle, on The Sequani Calendar of the Celts.

As representatives of the house cult of the Great Mother Goddess, the raven-goddesses are depicted in European iconography, especially in the Gaulish goddess of Nantosuelta, or "winding river," as a goddess often holding a dovecot or symbol of the house in one hand and a raven perched on her shoulder or in the company of a huge raven. In one icon, the goddess has the wings of the bird goddess (Ross 314). In the Celtic mythology of the British Isles, the raven-goddess is seen as more of a sovereign and protector of the tribe than her European counterparts. Like the Greek goddesses, Hera and Athena, the Celtic goddesses are protectors and sovereign figures. The Morrigan, in particular, acts as protector of the Tuatha Dé Danaan in *The Book of Conquests* and as protector of the hero, Cú Chulainn in the Ulster Cycle of Irish mythology.

In the heroic legends of the Celts, the hero, like the gods, must assume the powers of the bird and snake goddess by destroying her. This ambivalent relationship is depicted when Cú Chulainn meets the Morrigan at the ford. The Morrigan shape-changes to reveal many of the forms familiar to the Great Mother Goddess. She appears as a white, hornless, red-eared heifer with fifty heifers about her, an eel or watery serpent, a she-wolf, and a hag who must be healed by Cú Chulainn. Only Cú Chulainn has the powers to heal her and receive milk from the three-teated cow that the old hag is milking. This done, the Morrigan returns to her bird form (Ross 283). Aspects of the goddess such as her associations with milk, water, the cow, the serpent, and healing, as well as her bird attribute are represented. When Cú Chulainn destroys a flock of ravens, he performs a strange ritual with the last bird; "He cuts off its head and bathes his hands in its blood and sets the head on a rock which was then called Srub Brain, Raven's bill. Belief in the various powers of blood are well-known and widespread, and it was perhaps thought that actual combat with the blood of the magic otherworld raven would imbue the hero with some of its particular prophetic and supernatural powers" (Ross 325).

The raven-goddesses are powerful figures of divination in the mythology of the Celts. The Badv is "sometimes encountered at a ford in a river washing the armor and weapons of those who are about to die in battle;" her powers over the river are demonstrated when she makes the river bloody with gore in her divinations (Ross 314). The Badv, called "a royston crow or vulture" and "the blood-red-mouthed," terrifies with her shrieking as a reminder of the blood spilled in war (Dexter *Whence the Goddesses* 89). These black-raven goddesses who deprived their enemies of their strength through magic and incantation of doom were foreboding figures (Ross 315). According to Dexter, the warrior-goddess "took and gave blood, the stream of life. As did the Valkyries, she exhorted men on the field of battle without engaging in battle herself. And she bestowed prowess in battle through the transmission of sexual powers" (174). Her prophecies associated with sex and warfare are a potent reminder of the prophetic powers of the bird and snake goddesses of former eras.

In Druids' rituals, the raven is a prognostic bird that is much less ominous than the raven-goddesses. The Druids in Gaul and Ireland told the future by observing the flight of birds and the movements of the raven and the wren in particular. Many votive deposits, such as the flesh-fork from Dunaverney, have been found with prognostic equipment decorated with ravens and other birds. At Felmington Hall in Norfolk, little bronze ravens, human masks, and a wheel were found as part of votive deposits. According to Ross, there is ample evidence of the Druids' use of ravens in ritual and by domestication for the purpose of divination. Prophecy is also divined by understanding the "language of the birds" and holding conversation with them (Ross 327). Like Melampus in Greek mythology, the Druids believe that the language of the birds is the language of the goddess.

The owl and the eagle are both associated with the Neolithic mother goddess in Europe. Owls with bovine heads on a bronze cauldron from Jutland and owls on a gold torc from Reinheim as well as numerous birds from the La Tène toreutic works suggest a cult of some importance (Ross 345). Where the owl, or aderyn y corff, "corpse bird" in Welsh, and the raven are associated with blood and carnage reminiscent of the Neolithic excarnation rites, the eagle once again is associated with transformation and ascendance in Celtic mythology. In Irish mythology, Lugh, the Celtic god known as Lleu in Wales and Lugus on the continent, dies and turns into a great eagle that alights in a sacred oak. Lugh and the tree twined with mistletoe are represented on a silver cup from Lyon, France (Ross 346-49).

The height of the summer brings the festival of Lugnasad, a cross-quarter day, to the Celtic peoples where both the ascendance of the eagle, as Lugh, and the position of the Summer Triangle with its vultures, ravens, owls, swans, and finally Altair, the Rising One, in Aquila the Eagle, is set in the center of the night sky. The Milky Way and The Serpent Bearer and the Serpent as well as Draco, the Old Serpent coiling about the celestial north pole, create a spectacular image of the Great Mother Goddess of the summer months. Two full moons, marked as Tiocobrextio on The Sequani Calendar, complement the portrait of the goddess by representing the full moon harvests of herbs and vegetables from the garden of the Mother.

The celebration of Lugnasad is the center of the abundant harvest of the Great Mother Goddess and the celebration of the gifts she has bestowed upon the people. It contains the prophecy of the year to come from the abundance of the harvest. Disputes are settled, couples are wedded and the market fairs are blessed in remembrance of the sacrifice of the Mother Goddess. Celtic fairs and celebrations at Tailtinn or Telltown are in memory of Lugh's foster-mother, Tailtiu, who died clearing the fields and bringing in the harvest, and commitments are made for the upcoming year. In a sense, the year has come to an end and the worship of the Great Mother Goddess realized.

A statue of the Great Mother Goddess found in a peat bog at the spring of the Seine River outside of Dijon, France is one of the last and lasting images of the goddess. It is a statue of Sequana, the bird and river goddess of central Europe. Although her snakes are gone in this Gallo-Roman depiction of the Great Mother Goddess, her hands are open to embrace the people. She receives them and heals them through the miracle and milk of life, her sacred waters. It is through her worship that the Sequani tribe found the strength to preserve her calendar, an exquisite time-keeper for the stars, the moon, and the sun, in order that future generations may live life through its wonders. Time-keeper, earth-keeper, star-keeper for the people, we salute The Sequani Calendar and the Great Mother Goddess and all her glory.

Works Cited

Chapter One

Dow, Sterling and Robert F. Healey. *A Sacred Calendar of Eleusis: Harvard Theological Studies XXI*. Cambridge: Harvard UP, 1965.

Gimbutas, Marija. *The Living Goddesses*. Edited and Supplemented by Miriam Robbins Dexter. Berkeley/ Los Angeles: U of California P, 1999.

Kerenyi, Carl. *Eleusis: Archetypal Image of Mother and Daughter*. Princeton: Princeton UP, 1991.

Mallory, J.P. and D.Q. Adams "Priest." *Encyclopedia of Indo-European Culture*. Ed. J.P. Mallory and D.Q. Adams. London: Fitzroy Dearborn, 1997.

Mylonas, George E. *Eleusis and the Eleusinian Mysteries*. Princeton: Princeton UP, 1972.

Nilsson, Martin P. *The Minoan-Mycenaean Religion and Its Survival in Greek Religion*. New York: Biblo and Tannen, 1949.

Richardson, J.A. *Homeric Hymn to Demeter*. Oxford: Oxford UP, 1974.

Ross, Anne. *Pagan Celtic Britain*. Chicago: Academy Chicago Publishers, 1967.

Ruck, C.P., Gordon Wasson, and Albert Hoffman. *The Road to Eleusis*. Los Angeles: Hermes Press, 1998.

Chapter Two

Apollodorus. *Library and Epitome*. Ed. Sir James George Frazer. *Perseus Project*, Tufts University. 30 Oct. 2000 <http://www. perseus .tufts.edu/cgibin/ptext?lookup=Apollod.+2.4.>.

Dexter, Miriam Robbins. "The Hippomorphic Goddess and Her Offspring." *The Journal of Indo-European Studies*. 18 (1990): 286-307.

Ellis, Peter Berresford. *Dictionary of Celtic Mythology*. New York: Oxford UP, 1992.

Gantz, Jeffrey. Introduction. *The Mabinogion*. New York: Penguin, 1976.

Gimbutas, Marija. *The Living Goddesses*. Edited and Supplemented by Miriam Robbins Dexter. Berkeley: U of California P, 1999.

Kinsella, Thomas, ed. *The Tain from the Irish Epic Táin Bó Cuailnge*. Oxford: Oxford UP, 1970.

The Mabinogion. Trans. Jeffrey Gantz. New York: Penguin, 1976

Miller, Dean A., J. P. Mallory and D.Q. Adams. "Horse." *Encyclopedia Of Indo-European Culture*. Ed. J.P. Mallory and D.Q. Adams. London: Fitzroy Dearborn, 1997.

Ovid. "The Story of Perseus." Book IV *Metamorphosis*. Trans. Sir Samuel Garth, John Dryden et al. *The Internet Classics Archive*. Web Atomics, MIT, 1994. 30 Oct. 2000 <http://classics.mit.edu/ Ovid/ metam.4.fourth.html>.

Staal, Julius D.W. *The New Patterns in the Sky: Myths and Legends of the Stars*. Blacksburg, Virginia: McDonald and Woodward, 1988.

Chapter Three

Brú na Bóinne: Newgrange, Knowth, Dowth, and the River Boyne: A Supplement to Archaeology Ireland. 11.3 (1997).

Chartrand, Mark R. *National Audubon Society Field Guide to the Night Sky*. New York: Knopf, 1991.

Cornelius, Geoffrey and Paul Devereux. *The Secret Language of the Stars and Planets: A Visual Key to the Heavens*. San Francisco: Chronicle Books, 1996.

Dexter, Miriam Robbins. *Whence the Goddesses: A Source Book*. New York: Teachers College Press, 1990.

Dow, Sterling and Robert F. Healey. *A Sacred Calendar of Eleusis: Harvard Theological Studies XXI*. Cambridge: Harvard UP, 1965.

Eliade, Mircea. *The Myth of the Eternal Return*. Princeton, N.J.: Princeton UP, 1974.

Ellis, Peter Berresford. *Dictionary of Celtic Mythology*. New York: Oxford UP, 1992.

Gimbutas, Marija. *The Goddesses and Gods of Old Europe: Myths and Cult Images*. Berkeley: U of California P, 1996.

Graves, Robert. *The Greek Myths*. 2 Vols. New York: Penguin, 1960.
Griaule, Marcel. *Conversations with Ogotemmeli: An Introduction to Dogon Religious Ideas*. Oxford: Oxford UP, 1965.
Haudry, Jean. "The Cosmic Religion of the Indo-Europeans" *Languages and Cultures: Studies in Honor of Edgar C. Polomé.* Ed. Mohammad Ali Jazayery and Werner Winter. Berlin: Mouton de Gruyter, 1988.
The Mabinogion. Trans. Jeffrey Gantz. New York: Penguin, 1976
Murphy, Anthony and Richard Moore. "The Cygnus Enigma" *Mythical Ireland*. Ed. Anthony Murphy. 7 July 2001 <http://www.mythicalireland.com/cygnus/index.html>.
Nilsson, Martin P. *The History of Greek Religion*. New York: Biblo and Tannen, 1925.
O'Brien, Steven. "Divine Twins." *Encyclopedia of Indo-European Culture*. Ed. J. P. Mallory and D.Q. Adams. London: Fitzroy Dearborn, 1997.
Ross, Anne. *Pagan Celtic Britain*. Chicago: Academy Chicago Publishers, 1996.
Staal, Julius D.W. *The New Patterns in the Sky: Myths and Legends of the Stars*. Blacksburg, Va.: McDonald and Woodward, 1988.
Suzuki, David and Peter Knudtson. *Wisdom of the Elders: Sacred Native Stories of Nature*. New York: Bantam, 1993.

Chapter Four

Brennen, Martin. *The Stones of Time*. Rochester, Vermont: Inner Traditions, 1994.
Chartrand, Mark R. *National Audubon Society Field Guide to the Night Sky*. New York: Knopf, 1991.
Dexter, Miriam Robbins. *Whence the Goddesses: A Source Book*. New York: Teachers College Press, 1990.
Ellis, Peter Berresford. *Dictionary of Celtic Mythology*. New York: Oxford, UP, 1992.
Frazer, Sir James. *The New Golden Bough*. Theodor H. Gaster, Editor. NewYork: Signet, 1964.
Gimbutas, Marija. *The Goddesses and Gods of Old Europe: Myths and Cult Images*. Berkeley: U of California P, 1996.
——. *The Living Goddesses*. Edited and Supplemented by Miriam Robbins Dexter. Berkeley: U of California P, 1999.
Graves, Robert. *The Greek Myths*. 2 vols. New York: Penguin, 1960.

Griaule, Marcel. *Coversations with Ogotemmeli: An Introduction to Dogon Religious Ideas*. Oxford: Oxford UP, 1965: 108.

Kaul, Flemming and Ivan Marazov, Jan Best, Nanny De Vries. *Thracian Tales of the Gundestrap Cauldron*. Amsterdam: Najade Press, 1991.

Kerenyi, Carl. *Dionysus: Archetypal Images of Indestructible Life*. Princeton:Princeton UP, 1976.

Kinsella, Thomas, ed. *The Tain from the Irish Epic Táin Bó Cuailnge*. Oxford: Oxford UP, 1970.

The Mabinogion. Trans. Jeffrey Gantz. New York: Penguin, 1976

Miller, Dean A. "Threefold Death." *Encyclopedia of Indo-European Culture*. Ed. J.P. Mallory and D.Q. Adams. London: Fitzroy Dearborn, 1997.

——. "Warriors." *Encyclopedia of Indo-European Culture*. Ed. J.P. Mallory and D.Q. Adams. London: Fitzroy Dearborn, 1997.

Mylonas, George. *Eleusis and the Eleusinian Mysteries*. Princeton: Princeton UP, 1972.

Nilsson, Martin P. *The Dionysiac Mysteries of the Hellenistic and Roman Age*. New York: Arno Press, 1975.

Otto, Walter F. *Dionysus: Myth and Cult*. Bloomington, Indiana: Indiana UP, 1965.

Ross, Anne. *Pagan Celtic Britain*. Chicago: Academy Chicago Publishers, 1996.

——. *Druids*. London: Tempus, 1999.

Chapter Five

Brennan, Martin. *The Stars and the Stones: Ancient Art and Astronomy in Ireland*. London: Thames and Hudson, 1983.

Chartrand, Mark R. *National Audubon Society Field Guide to the Night Sky*. New York: Knopf, 1991.

Cornelius, Geoffrey and Paul Devereux. *The Secret Language of the Stars and Planets: A Visual Key to the Heavens*. San Francisco: Chronicle Books, 1996.

Dexter, Miriam Robbins. *Whence the Goddesses: A Source Book*. New York: Teachers College Press, 1990.

——. "Reflections on the Goddess *Donu." *The Mankind Quarterly* 30 (1990): 45-58.

Ellis, Peter Berresford. *Dictionary of Celtic Mythology*. New York: Oxford, 1992.

Gimbutas, Marija. *The Goddesses and Gods of Old Europe: Myths and Cult Images*. Berkeley: U of California P, 1996.

——. *The Living Goddesses*. Ed. Miriam Robbins Dexter. Berkeley/Los Angeles: U of Calfornia P, 1999.

Graves, Robert. *The Greek Myths*. 2 vols. New York: Penguin, 1960.

Griaule, Marcel. *Conversations with Ogotemmeli: An Introduction to Dogon Religious Ideas*. Oxford: Oxford UP, 1965.

Hancock, Graham and Santha Faiia. *Heaven's Mirror: Quest for the Lost Civilization*. New York: Three Rivers Press, 1998.

Heaney, Marie. *Over Nine Waves*. London: Faber and Faber, 1995.

Hopman, Ellen Evert. *A Druid's Herbal for the Sacred Earth Year*. Rochester, Vermont: Destiny Books, 1995.

Neumann, Erich. *The Origins and History of Consciousness*. Bollingen Series XLII. Princeton: Princeton UP, 1973.

Nilsson, Martin. *The Minoan-Mycenaean Religion and Its Survival in Greek Religion*. New York: Biblo and Tannen, 1949.

Ross, Anne. *Pagan Celtic Britain*. Chicago: Academy Chicago Publishers, 1996.

Staal, Julius D.W. *New Patterns in the Sky: Myths and Legends of the Stars*. Blacksburg, VA: McDonald and Woodward, 1988.

Bibliography

Apollodorus. *Library and Epitome*. Ed. Sir James George Frazer. *Perseus Project*, Tufts University. 30 Oct. 2000 <http://www.perseus.tufts.edu/cgibin/ptext?lookup=Apollod.+2.4.>.

Biaggi, Cristina. *Habitations of the Great Goddess*. Manchester, CT: Knowledge, Ideas & Trends, 1994.

Brennan, Martin. *The Stars and the Stones: Ancient Art and Astronomy in Ireland*. London: Thames and Hudson, 1983.

——. *The Stones of Time*. Rochester, Vermont: Inner Traditions, 1994.

Brú na Bóinne: Newgrange, Knowth, Dowth, and the River Boyne: A Supplement to Archeology Ireland. 11.3 (1997).

Campbell, Joseph. *The Hero with a Thousand Faces*. New York: Pantheon, 1973.

——. *The Masks of God*. New York: Viking Press, 1962.

——. *Transformations of Myth Through Time*. New York: Perennial Library, 1990.

Chartrand, Mark R. *National Audubon Society Field Guide to the Night Sky*. New York: Knopf, 1991.

Cornelius, Geoffrey and Paul Devereux. *The Secret Language of the Stars And Planets: A Visual Key to the Heavens*. San Francisco: ChronicleBooks, 1996.

Dexter, Miriam Robbins. *Whence the Goddesses: A Source Book*. New York: Teachers College Press, 1990.

——. "The Hippomorphic Goddess and Her Offspring" *The Journal of Indo-European Studies*. 18 (1990): 286-307.

——. "Reflections on the Goddess *Donu" *The Mankind Quarterly*. 30 (1990): 45-58.

Dow, Sterling and Robert F. Healey. *A Sacred Calendar of Eleusis: Harvard Theological Studies XXI*. Cambridge: Harvard UP, 1965.

Eliade, Mircea. *The Myth of the Eternal Return*. Princeton: Princeton UP, 1974.

Ellis, Peter Berresford. *Dictionary of Celtic Mythology*. New York: Oxford UP, 1992.

Frazer, Sir James. *The New Golden Bough*. Theodore H. Gaster, Editor. New York: Signet, 1964.

Gantz, Jeffrey. Introduction. *The Mabinogion*. New York: Penguin, 1976.

Gimbutas, Marija. *The Goddesses and Gods of Old Europe: Myths and Cult Images*. Berkeley: U of California P, 1996.

——. *The Living Goddesses*. Ed. Miriam Robbins Dexter. Berkeley: U of California P, 1999.

Graves, Robert. *The Greek Myths*. 2 vols. New York: Penguin, 1960.

Griaule, Marcel. *Conversations with Ogotemmeli: An Introduction to Dogon Religious Ideas*. Oxford UP, 1965.

Hancock, Graham and Santha Faiia. *Heaven's Mirror: Quest for the Lost Civilization*. New York: Three Rivers Press, 1998.

Haudry, Jean. "The Cosmic Religion of the Indo-Europeans" *Languages And Cultures: Studies in Honor of Edgar C. Polome*, Ed. Mohammad Ali Jazayery and Werner Winter. Berlin: Mouton de Gruyter, 1988.

Heaney, Marie. *Over Nine Waves*. London: Faber and Faber, 1995.

Hopman, Ellen Evert. *A Druid's Herbal for the Sacred Earth Year*. Rochester, Vermont: Destiny Books, 1995.

Jenkins, John Major. *Galactic Alignment: The Transformation of Consciousness According to Maya, Egyptian and Vedic Traditions*. Rochester, VT: Inner Traditions, 2002.

——. *Maya Cosmogenesis 2012: The True Meaning of the Maya Calendar End-Date*. Santa Fe, NM: Bear & Company, 1998.

Jung, Carl Gustav. *Four Archetypes: Mother, Rebirth, Spirit, Trickster*. Princeton: Princeton UP, 1970.

——. *Man and His Symbols*. Garden City, NY: Doubleday, 1964.

Kaul, Flemming and Ivan Marazov, Jan Best, Nanny De Vries. *Thracian Tales of the Gundestrap Cauldron*. Amsterdam: Najade Press, 1991.

Kerenyi, Carl. *Dionysus: Archetypal Images of Indestructible Life*. Princeton: Princeton UP, 1976.

——. *Eleusis: Archetypal Image of Mother and Daughter*. Princeton: Princeton UP, 1991.

Kinsella, Thomas, ed. *The Tain from the Irish Epic Táin Bó Cuailnge*. Oxford: Oxford UP, 1970.

Leeming, David Adams. *A Dictionary of Creation Myths*. New York: Oxford UP, 1996.

——. *Mythology: The Voyage of the Hero*. New York: Oxford UP, 1998.

——. *The World of Myth*. New York: Oxford UP, 1992.

The Mabinogion. Trans. Jeffrey Gantz. New York: Penguin, 1976

Mallory, J.P. and D.Q. Adams. "Priest" *Encyclopedia of Indo-European Culture*. Ed. J.P. Mallory and D.Q. Adams. London: Fitzroy Dearborn, 1997.

McKenna, Terence. *The Archaic Revival: Speculations on Psychedelic Mushrooms, the Amazon, Virtual Reality, UFOs, Evolution, Shamanism, the Rebirth of the Goddess, and the End of History*. New York: HarperCollins, 1991.

——. *Food of the Gods: The Search for the Original Tree of Knowledge: A Radical History of Plants, Drugs and Human Evolution*. New York: Bantam, 1992.

Michell, John F. *Megalithomania*. Ithaca: Cornell UP, 1982.

——. *Twelve-Tribe Nations and the Science of Enchanting the Landscape*. Grand Rapids: Phanes Press, 1991.

Miller, Dean A. "Threefold Death." *Encyclopedia of Indo-European Culture*. Ed. J.P. Mallory and D.Q. Adams. London: Fitzroy Dearborn, 1997.

——. "Warriors" *Encyclopedia of Indo-European Culture*. Ed. J.P. Mallory and D.Q. Adams. London: Fitzroy and Dearborn, 1997.

Miller, Dean A., J.P. Mallory, and D.Q. Adams. "Horse" *Encyclopedia of Indo-European Culture*. Ed. J.P. Mallory and D.Q. Adams. London: Fitzroy Dearborn, 1997.

Murphy, Anthony and Richard Moore. "The Cygnus Enigma." *Mythical Ireland*. Ed. Anthony Murphy. 7 July 2001 <http://www.mythicalireland.com/cygnus>.

Mylonas, George. *Eleusis and the Eleusinian Mysteries*. Princeton: Princeton UP, 1972.

Neumann, Erich. *Art and the Creative Unconscious: Four Essays*. New York: Harper and Row, 1966.

——. *The Great Mother: Analysis of the Archetype*. New York: Pantheon, 1963.

——. *The Origins and History of Consciousness*. Bollingen Series XLII. Princeton: Princeton UP, 1973.

Nilsson, Martin P. *The History of Greek Religion*. New York: Biblo and Tannen, 1925.

——. *The Minoan-Mycenaean Religion and Its Survival in Greek Religion*. New York: Biblo and Tannen, 1949.

——. *The Dionysiac Mysteries of the Hellenistic and Roman Age*. New York: Arno Press, 1975.

O'Brien, Steven. "Divine Twins." *Encyclopedia of Indo-European Culture*. Ed. J.P. Mallory and D.Q. Adams. London: Fitzroy Dearborn, 1997.

Otto, Walter F. *Dionysus: Myth and Cult*. Bloomington, Indiana: Indiana UP, 1965.

Ovid. "The Story of Perseus." Book IV *Metamorphosis*. Trans. Sir Samuel Garth, John Dryden et al. *The Internet Classics Archive*. Web Atomics, MIT, 1994. 30 Oct. 2000 <http://classics.mit.edu/Ovid/metam.4.fourth.html>.

Richardson, J.A. *Homeric Hymn to Demeter*. Oxford: Oxford UP, 1974.

Ross, Anne. *Pagan Celtic Britain*. Chicago: Academy Chicago Publishers, 1996.

——. *Druids*. London: Tempus, 1999.

Ruck, C.P., Gordon Wasson, and Albert Hoffman. *The Road to Eleusis*. Los Angeles: Hermes Press, 1998.

Staal, Julius D.W. *New Patterns in the Sky: Myths and Legends of the Stars*. Blacksburg, VA: McDonald and Woodward, 1988.

Suzuki, David and Peter Knudtson. *Wisdom of the Elders: Sacred Native Stories of Nature*. New York: Bantam, 1993.

Index

www.ingramcontent.com/pod-product-compliance
Lightning Source LLC
LaVergne TN
LVHW090411210926
848480LV00002B/525